Thomas Addison

A Collection of the Published Writings of the Late Thomas Addison,

M.D., physician to Guy's Hospital

Thomas Addison

A Collection of the Published Writings of the Late Thomas Addison, M.D., physician to Guy's Hospital

ISBN/EAN: 9783337161217

Printed in Europe, USA, Canada, Australia, Japan

Cover: Foto ©berggeist007 / pixelio.de

More available books at **www.hansebooks.com**

A COLLECTION

OF

THE PUBLISHED WRITINGS

OF THE LATE

THOMAS ADDISON, M.D.,

PHYSICIAN TO GUY'S HOSPITAL.

EDITED, WITH INTRODUCTORY PREFACES TO SEVERAL OF THE PAPERS,

BY

DR. WILKS AND DR. DALDY.

THE NEW SYDENHAM SOCIETY,
LONDON.

MDCCCLXVIII.

CONTENTS.

	PAGE
BIOGRAPHY	ix
I. OBSERVATIONS ON THE ANATOMY OF THE LUNGS . . .	1
II. OBSERVATIONS ON THE DIAGNOSIS OF PNEUMONIA . .	7
III. OBSERVATIONS ON PNEUMONIA AND ITS CONSEQUENCES . .	17
IV. ON THE PATHOLOGY OF PHTHISIS	39
V. ON THE DIFFICULTIES AND FALLACIES ATTENDING PHYSICAL DIAGNOSIS IN DISEASES OF THE CHEST	65
VI. OBSERVATIONS ON FATTY DEGENERATION OF THE LIVER . .	99
VII. ON THE DISORDERS OF FEMALES CONNECTED WITH UTERINE IRRITATION	109
VIII. CASE OF OVARIAN DROPSY REMOVED BY THE ACCIDENTAL RUPTURE OF THE CYST	155
IX. ON A CERTAIN AFFECTION OF THE SKIN, VITILIGOIDEA—α. PLANA, β. TUBEROSA	157
X. ON THE KELOID OF ALIBERT AND ON TRUE KELOID . .	165
XI. ON THE DISORDERS OF THE BRAIN CONNECTED WITH DISEASED KIDNEYS	187
XII. ON THE INFLUENCE OF ELECTRICITY AS A REMEDY IN CERTAIN CONVULSIVE AND SPASMODIC DISEASES . . .	195
XIII. ON THE CONSTITUTIONAL AND LOCAL EFFECTS OF DISEASE OF THE SUPRA-RENAL CAPSULES	211
INDEX	241

BIOGRAPHY.

THE records of Addison's early life are so difficult of ascertainment that any account of his actual life, in reference to the object of these papers, must date from the commencement of his association with Guy's Hospital about the year 1819 or 1820. That must have been nearly the period at which he attached himself to the hospital, where his enterprising and spirited activity in the search after a definite explanation of every form of disease presented to his observation, attracted the attention of the observant and large-minded treasurer; who was then by common acknowledgment the great *administrative* benefactor, and, through the appreciation of his minute, vigorous, and just guidance of its business from the smallest details to the highest principles involved in its government, the *accepted* dictator of the affairs of the institution. It was not easy in those days to overcome the prevailing *ingrained* prejudice that unless a man had been originally a pupil of the hospital he was not fairly eligible to the duties of its offices. Addison's may be cited, indeed, as one of the earliest (if not the first) instances of those traditional trammels being broken through; for in 1824 he was appointed assistant-physician to the hospital, his previous association with it consisting only in his entry there as a student after having taken his degree at Edinburgh.

It cannot be doubted that Mr. Harrison's intelligent

and scrutinising attention to the qualities of the gentlemen then *about* the hospital in reference to the appointments to future vacancies induced him to select Addison as a physician highly calculated to carry out the two foremost objects of his life, the promotion of the usefulness of the charity, and the extension of the reputation of the school.

There is always an interest attaching to the history of any man who has made a prominent position for himself in whatever career of life; the question " how he did it" excites the interest of the younger members of any profession in which it has been achieved. In the instance of Addison the old answer must be given —he " did it" by indomitable perseverance in the pursuit of *one* object of study; making it, as it were, the one day-dream and night-dream of his existence.

He seems to have been born of humble parents at Long Benton, near Newcastle, about 1793, to have been sent to the village school there, and afterwards to the Grammar School at Newcastle. Thence he migrated to Edinburgh, where he graduated M.D. in 1815, selecting for the subject of his inaugural thesis " De Siphilide."

His familiarity with Latin was at this time exemplified by a habit of taking down lecture-notes in that language; and this mental peculiarity may have led to his habitual exactitude of diction in whatever he wrote or spoke in after life.

On his arrival in London, he took up his residence at Skinner Street, Snow Hill, in one of the so-called " haunted houses" (possibly from some association with the Cock Lane ghost story of Dr. Johnson), and was soon after appointed house-surgeon to the Lock Hospital, where he acquired so great an interest in the subject of syphilis, that, although a topic not in strict accord with the branch of the profession which he had adopted, he always spoke on it authoritatively.

The next step in his career which we can gather is his residence in Hatton Garden, and his attachment to the General Dispensary. At this time he studied with the celebrated Bateman, and thus became so great a proficient in skin-disease as to acquire the acknowledgment of those cognisant of his skill, that after the death of the great dermatologist his mantle had fallen upon Addison. From his desire to avoid the pursuit of this subject as a specialty, this competency was not generally known; but it is certain that until a comparatively recent period Addison's accuracy in the discrimination of cutaneous eruptions was scarcely matched; an assertion which a careful examination of the unrivalled wax models in the museum of Guy's Hospital (studiously prepared during many years under his own superintendence) will fully bear out. It is also believed that he entered as a student at St. George's Hospital, but of this we can obtain no proof, and know only that he became successively a student at Guy's, was promoted early to the office of assistant-physician (1824), when, from the recognition of his great practical knowledge, his fame spread rapidly among the pupils, and he became a brilliant acquisition to the new school. This recognition placed him, in consequence of an early vacancy (1827), in the chair of Materia Medica. At that time, when medical students paid fees for separate courses of lectures, they sought throughout the metropolis for the most attractive teachers. Armstrong then drew a large class to the Webb Street school by his instruction in the practice of medicine; most of his pupils remained to listen to

Addison, and so great was the attendance that his lecture-fees must have amounted to £700 or £800 a year.

In the year 1837 he received the appointment of physician to the hospital, and was at the same time selected as the colleague of the late Dr. Bright in the duties of the chair of Medicine.

At this period he commenced, conjointly with his colleague, a work on 'Medicine.' From the high estimation in which this work was held it must be a matter of regret that one volume only was published. Now that both these widely-known authors have departed from their labours, it cannot be harmful to assert (what was then generally known) that the greater portion of the work was from the pen of Dr. Addison.

He had now achieved the desired position for the development, or rather the showing forth, of the qualities which he had cultivated with so much care; those of the eminently *practical* physician. And he certainly exhibited them in a remarkable degree; his strong, positive, and perpetual insistance upon the term "*practical*," in reference to disease, constitutes, indeed, the key to Addison's character and professional career. He was always ready to discuss newly-started theories, but he never, for a moment, allowed them to interfere with the results of his matured experience. Possessing unusually vigorous perceptive powers, being shrewd and sagacious beyond the average of men, the patient before him was scanned with a penetrating glance from which few diseases could escape detection. He never reasoned from a half-discovered fact, but would remain at the bedside, with a dogged determination to track out the disease to its *very* source for a period which constantly wearied his class and his attendant friends.

So severely did he tax his mind with the minutest details bearing upon the exact exposition of a case, that he has been known to startle the sister of the ward in the middle of the night by his presence; after going to bed with the case present to his mind, some point of what he considered important detail in reference to it occurred to him, and he could not rest until he had cleared it up. He has also been known after seeing a patient within the radius of eight or ten miles to have remembered on his near approach to London, thinking over the case on his way, that he had omitted some seemingly important inquiry, and to have posted back some miles for the purpose of satisfying his mind on the doubt which had occurred to it. If at last he could lay his finger on the disease, his victory was attained, and his painstaking satisfactorily rewarded. For with him accurate "diagnosis" was the great, and too often the *ultimate*, object of an industry of search, a correlation of facts deduced from *scientific* observation, and a concentration of thought rarely combined in the individual physician.' To those who knew him best his power of searching into the complex framework of the body, and dragging the hidden malady to light, appeared unrivalled; but we fear that the *one* great object being accomplished, the same energetic power was not devoted to its alleviation or cure. Without accusing Addison of a meditated neglect of therapeutics, we fancy that we can trace the dallying with remedies which has been the characteristic of more recent times. " I have worked out the disease; if it be remediable, Nature, *with fair play*, will remedy it. I do not clearly see my way to the direct agency of special medicaments, but I must prescribe something for the patient, at least, to satisfy his or her friends," seems to have been a part of the habit of mind which can deal satisfactorily only with the "*observable and proven,*" and shrinks from the "*uncertain and questionable.*"

His discovery of the heretofore unsuspected disease of the supra-renal capsules was the result of an exhaustive analysis of every organ of the body, without elucidating any evident reason for a remarkable form of anæmia: for a time he was constrained to the expression "idiopathic anæmia," accompanied by a prognosis of its fatal issue. This prognosis was so constantly verified that, following up the doctrine of exclusion, he at last, in the absence of any other noticeable cause for it, observed an association between it and a peculiar appearance of the supra-renal capsules. Here we trace the advantage of the *large* study of clinical facts; at that time skin disease was supposed to be confined to the province of the surgeon, and it is probable that but for Addison's accurate observation of the various cutaneous deviation from its ordinary and healthy condition, the bronzed skin would not have rivetted his attention so forcibly as to have incited him to prosecute his inquiries to their ultimate issue.

We recognise the necessity for some brief remarks on Addison's *disposition* from the conviction that it was not generally understood. Viewed in its professional aspect, no character on record has presented in a higher degree the sterling hard qualities of true professional honesty. We have never heard a single instance in which a word of disparagement against a professional brother escaped him. He would always strenuously and with all his natural vigour maintain what he believed to be the truth, but never for the purpose of underrating the opinions of others. His whole bearing in the profession was to the last degree honorable, and anything like jealousy or ill-will against another professional man never entered his mind.

The admirable bust by Towne, which now adorns the museum of Guy's Hospital, is the best exposition of the estimation in which he was held by his colleagues, from whose subscriptions it resulted.

He was for many years acknowledged as the spirit which influenced the medical doings at Guy's Hospital, and to Addison is due, in great measure, the prominent character which the medical department of that institution has of late years held in public professional estimation.

Yet in professional intercourse his disposition presented peculiarities often misinterpreted by the professional observer: he, the observer, saw what appeared to him a rudeness, a certain bluntness of expression conveying to him the idea of a haughtiness, or at least of Addison's assumption of superiority; so that he parted with him impressed with the dignity of his bearing, a full appreciation of the accurate and well-sifted opinion which he had obtained, but at the same time carrying with him the notion that, judging from his apparently unapproachable manner, and what seemed to many "hauteur," he was a man of large self-esteem. This is one of the commonest mistakes in the estimation of character. In what degree a resoluteness of expression and an undue energy of manner is unconsciously adopted to cloak a covert *physical* nervousness, no one but the wearer of the cloak can fully estimate. We have reason to know that Addison suffered most acutely from this physical enervation; he has even said, "I never rose to address the Guy's *Junior* Physical Society without feeling nervous," and yet his listeners would depart with the feeling that he had spoken to them in what may almost be termed a tone of "bluster;" they went away impressed with the dignity of his bearing, and crediting him with great physical and moral energy; not recognising that a quick, hasty, and impassioned manner of expression is not infrequently the result of a deficient controlling power. We know that his mind was to the last degree susceptible, and that although wearing the outward garb of resolution, he was, beyond most other men, most liable to sink under trial. We lay some stress upon this peculiarity for the purpose of vindicating his character from the unamiable spirit which we have heard sometimes laid to his charge. If there be any of our readers who may have been vexed by the apparent discourteousness of his manner, let them carefully consider the explanation of it which we have attempted to depict, and give thanks to God that they have been blessed with a calmer and less perturbable spirit.

His last communication addressed to the pupils of the hospital in answer to a letter of condolence written to him on his retirement from the hospital on account of his ill-health conveys most markedly the seemingly latent kindliness of his nature, his entire devotion to whatever would aid to place medicine on a more scientific basis, his affectionate regard for those of his juniors who were fellow-helpers in the work, and his own powerful style of expressing what he sincerely and heartily felt.

"*March* 17*th*, 1860.

"MY DEAR SIR,

"A considerable break-down in my health has scared me from the anxieties, responsibilities, and excitement of the profession: whether temporarily or permanently cannot yet be determined; but, whatever may be the issue, be assured that nothing was better calculated to soothe me than the kind interest manifested by the pupils of Guy's Hospital during the many trying years devoted to that institution.

"I can truly affirm that I ever found my best support and encouragement in the generous gratitude and affectionate attachment, as well as my proudest reflections, in the honorable and most exemplary conduct of its pupils. Present my sincere regards and best wishes to every one of them, and believe me,

"Yours truly and affectionately,

"Thomas Addison."

"E. Galton, Esq."

PREFACE

TO THE FIVE FOLLOWING PAPERS, RELATING TO

DISEASES OF THE LUNGS.

ONE of the conspicuous features of medical literature is the amount of valuable information which it contains in a scattered and disjointed form.

Two of the prominent causes of this result are the honest modesty which induces some men to shrink from publication in a concrete form, under the sincere belief that others are as well informed on their special subject of pursuit as themselves, and the lack of industry in many to gather up what they know in a succinct and constructed shape.

The prior cause or fault (if fault it be) must be imputed to Addison; for as one of the early expositors of, and progressionists in, the teachings of Laennec, Addison held a foremost position in this country: yet we find him more than thirty years since reading papers on the subject of *lung disease*, of which the importance and great value seem not yet to be fully estimated, to the Physical Society of Guy's Hospital and the South London Medical Society. These papers would, we believe, under the more ready diffusion of information adopted at the present time, have then attracted the attention of the most learned physicians of Europe, and would thus have contributed to an earlier knowledge of the pathology of pneumonia and phthisis.

With lack of industry and constructive arrangement Addison certainly cannot be reproached; for his conclusions were deduced from the most accurate necroscopic inquiries, pursued, not only at the well-directed post-mortem examinations of Guy's Hospital, but at his own residence also, with an assiduousness and a sedulous perseverance comparable to the zeal of a Harvey or a Hunter; and his conclusions are expressed in a terse, forcible, and readily intelligible language similar to theirs.

The main ground for the belief of the Council of the New Sydenham Society that a collection of Addison's papers would be acceptable to its members originated in the impression that the views of Addison on pneumonia and phthisis, taught during so many years, had not yet received due consideration: for even now the term "phthisis," implying simply a state of lung prone to disorganization, is regarded by too many of our profession as having always a basis of tubercle; and a consequent uniform mode of treatment is adopted, having commonly for its object a combat with this imaginary deposit.

For the fulfilment of the Society's object it seemed to the editors necessary either to republish what

Addison had written on diseases of the lung and their diagnosis only, which would have formed a small but incomplete volume; or to include all the separate papers which he had written, under the conviction that all he wrote was the outcome of sterling honest reflection upon carefully deduced facts : and that admitting, as they distinctly do, the various degrees of merit in his several papers, their publication, as a whole, would be advantageous to the members.

It must be remembered that Addison's influence, as an instructive physician, was great rather by his oral teachings as a lecturer than by his contributions to medical literature ; and that in the latter sphere essays, having great value and originality at the time of their production, have in the course of a quarter of a century lost much of their initiative merit by subsequent absorption into general medical literature. Although his fame may for ever rest on the disease which bears his name, we may remind our readers that his reputation as a teacher had been established long before the publication of his monograph on disease of the suprarenal capsules at the close of his career.

Foremost amongst his oral and written teachings were those which related to "*diseases of the lungs.*" He had insisted from the outset of his public professional life that the then taught doctrine concerning the condition of the lung in pneumonia was erroneous, and based upon a false view of the anatomy of the organ. He first controverted the opinion that pneumonia depended upon an inflammation of the then called "*parenchyma*" of the organ (*i. e.*, of a supposed areolar tissue connecting the cells, *arranged* according to the then adopted ideas of Riessessen) by denying that any such tissue existed except between the lobules.

In the first paper which we transcribe, read before the Medico-Chirurgical Society (1840, and taught in his lectures some years before this publication), he shows that there is no such connective tissue between the air-cells, as Riessessen conceived; that Reissessen's notion of every cell having a separate communication with a capillary bronchial tube was incorrect, and that consequently the inflammatory material exuded under the process of pneumonia could obtain no other localisation than in the air-cells themselves.

As the object of this publication is to inculcate peculiarly the doctrines of Addison, we may be allowed to lay some stress upon this incipient truth, from which so much of his ultimate deduction proceeded. If a comparison be made between his expressed opinions and those of both contemporary and subsequent writers, his inception will stand out more prominently.

For instance, so late as February, 1856, a reviewer of Dr. Barlow's 'Manual of the Practice of Medicine,' in which he reproduces Addison's ideas, writes, "Dr. Barlow defines pneumonia to be *inflammation of the air-cells of the lungs*," a definition not only calculated to give erroneous ideas to the student, but certainly not correct. True pneumonia nearly all authorities we thought were now agreed in regarding as an inflammation of the *interstitial tissue*, or parenchyma of the lungs: a view all but definitely settled by the experiments of Gendrin. It may, it is true, originate in the air-cells, but it may also have its *point-de-depart* in the mucous membranes of the bronchial tubes or in the pleura."

In 1843, Addison writes, " There are probably some present who remember the time and occasion when, in this Society and in opposition to all existing authorities, I ventured to call in question the long-cherished notion that pneumonia had its seat in a supposed parenchyma of the lungs, and that the products of pneumonic inflammation were poured into that parenchyma. Since that time I have had the satisfaction of witnessing a gradual but comparatively rapid renunciation of the latter views, and the adoption of those advanced in this Society so many years ago, viz., that pneumonia has its original and essential seat in the air-cells of the lungs, and that the ordinary pneumonic deposits are poured into these cells. It is nevertheless true that some of our most recent authorities are opposed to this opinion, and maintain that the pneumonic deposits are poured into an interstitial tissue; a conclusion which I find myself unable to reconcile with either the healthy or the morbid anatomy of the lungs."

It would thus appear that Addison was the suggestor of our present notions of the pathology of pneumonia.

In the second paper, which we reprint, " On the Diagnosis of Pneumonia," it may be remarked that at the time when it was written the stethoscope was not in general use; and that Addison, therefore, addressed his observations both to those who placed a full confidence in their recognition of the *general* signs of its existence, without reference to the *local or physical signs;* and to those who, enchanted with this new physical aid to diagnosis, relied entirely on its indications, ignoring the general signs which had guided many of their acutely observant predecessors into an accuracy of differential opinion in lung disease which often excited the admiration, and even wonder, of the early auscultators.

It would seem that in his earliest teaching and early writing his dominant idea was that a simple inflammatory deposit in the lung-cells or interlobular tissue was often mistaken for tubercle; he next observed that the *characteristic pungent heat of the skin* was an almost pathognomonic sign of pneumonia, which has been quoted and adopted by most succeeding authors; and that the pain which usually accompanies pneumonia was present only when pleurisy co-existed with it.

In the third paper, entitled "Observations on Pneumonia, and its consequences," after discussing the first stage of pneumonia, and admitting the analogy between an early stage of inflammation of the air-cells and that of serous membranes generally, *an analogy taught about that time by Dr. Stokes in the Dublin school,* he proceeds to illustrate the difference between red and grey hepatization. In the former, he shows that the great vascularity of the basement membrane and deterioration of the cell-walls produce consolidation by a jamming together, as it were, of the interior surfaces of the cells without any effusion into them; and that in the latter an albuminoid effusion into the cells, varying in its quality with the constitutional state of the patient, is the cause of the consolidation, and that this solid effusion may present post-mortem the appearance of—1st, *the uniform albuminous effusion;* 2nd, *the granular induration;* and, 3rd, *the grey induration.* These conditions of the lung, he contends, may in the course of their various stages of disintegration present many of the aspects commonly confounded with those of *tubercular phthisis.*

So important did this part of their subject appear to the editors, that they have thought it desirable to devote that portion of the limited amount which the funds of the Society permitted them to employ in *coloured* illustration to its especial exemplification.

The amplification of this conception will be found fully treated of in the fourth paper on " The Pathology of Phthisis," a paper judged by us to be the most important of his writings, because he therein matures his long-cherished notions concerning the chronic affections of the lungs recognised under the name of " Phthisis." In the dead-house this was for years the one absorbing subject of his investigation, as well as of his subsequent cogitation.

Addison declared that in the days of his early professional life it was not unusual to hear medical men express themselves, even in cases presenting the gravest pulmonary symptoms, hopeful of the issue, because there was as yet no proof that tubercles were formed; and that even when a fatal issue was evident the presence of tubercle was urged in reference to the question of prognosis.

With these impressions Addison devoted himself enthusiastically to unravel *pathologically* what appeared to him *clinically* inexplicable; and in the course of his investigations he so frequently found the deposit resulting from inflammation confounded with tubercle that he came gradually to look on the term "tubercular" in reference to what was customarily called "phthisis," as a bugbear obstructing the true and scientific explanation of chronic disorganization of the lung.

During the whole of the period that he occupied the chair of medicine at Guy's Hospital, he endeavoured to combat the prevailing idea that tubercle was the sole deadly ingredient in chronic pulmonary disease; and constantly asserted that in a large proportion of cases in which persons died of so-called "phthisis" tubercles were not present. He thought that the material deposited in the lung with a tendency to break up is often of an inflammatory kind, and may be produced under a variety of conditions.

He allowed, of course, that in a certain, perhaps large, number of cases tubercles were present, and that in persons of tuberculous constitution the formation of lowly-organized deposits occurred; but he held, at the same time, that similar deposits might be produced (in the absence of tubercle) under a great variety of conditions, especially under different forms of cachexia; or that an attack of pneumonia might proceed to disorganization, and be attended by all the symptoms usually accompanying "phthisis;" but he would not allow such a case to be ranked in the same category with a purely tuberculous affection. The most striking example of this idea is to be found in the instance of a case of pneumonic phthisis illustrated by Plate 7 attached to this paper. Here is shown a complete disintegration of the lung taken from a person who died of diabetes mellitus without a trace of tubercle. In this case, as in many others referred to, there is no corroboration of the idea that the adventitious deposit in the lung is " *tubercle*" by the discovery of it in any other organ; a fact not lightly assumed, but resulting from most *careful* necroscopic investigation, which lends great force to Addison's teaching.

We believe that if funds and space would have admitted the publication of the numerous drawings deposited in the museum of Guy's Hospital, which Addison had made from specimens prepared by himself, to illustrate the various forms of disease grouped under the names of "*phthisis and pneumonia*," no more valuable contribution to the knowledge of these subjects could have been offered to the profession.

These drawings exhibit almost every variety of chronic pulmonary disease which is met with; and it may repay any member of the New Sydenham Society who is interested in the subject to study them with care. He will be prepared, after our remarks, to find that in Addison's mind the subjects of pneumonia and phthisis were inseparably connected, and that when for the purpose of nosological arrangement he treats of the one, the other is inseparably blended with it.

At page 10, Dr. Addison writes: " It has been observed that these deposits (albuminous) may remain passive for an unlimited period, and without undergoing any very appreciable change; except, perhaps, a conversion of some of them into calcareous or chalky masses, especially when deposited in the upper lobe of the lung. It would nevertheless appear that the vital influence by which they are maintained in their integrity is so extremely slender that if inflammation happen to be set up around them by an accidental cause, and especially if the vital powers of the patient have been greatly impaired, that influence is so far exhausted that they lose their cohesion and soften." Although in these papers we find no definite application of the same conception to the instance of miliary tubercle, Addison orally taught with some emphasis that this form of tubercle might exist in a quiescent state for an indefinite time without serious apprehension on the part of the friends or the medical practitioner (the suspicion of its existence being induced only by dry, frequent irritable cough and uniform feeble respiratory murmur over the whole extent of the pulmonary walls, and a defective nutrition of the tissues of the body generally), until inflammation was set up in the cells where the tubercles were located, and that then the lungs disintegrated so rapidly and extensively as to give rise to the expression " galloping consumption." *

* The only occasion on which I ever met the late Dr. Chambers in consultation was a case of this kind, in 1848. I represented to him that the lady, æt. 34, had been the victim of miliary tubercle for certainly four years, that within the last month inflammation had been set up around the tubercles, and that she was dying rapidly of lung disorganization, and that I had eighteen months since forewarned her friends of this probable result. After examination of the patient, Dr. Chambers said to me, "Well, sir, I can only tell you that you are more fortunate

This subject is incidentally alluded to at page 79, proposition 23.

The forty-two " propositions " laid down by Addison in his paper on the " Difficulties and Fallacies attending Physical Diagnosis in Diseases of the Chest," are as true, as cogent, as applicable to the emergencies of daily practice *now*, as when they were published twenty-two years ago. Although many of the " fallacies " therein described have become familiar to, and consequently avoided by, the experienced stethoscopist, his axioms must be ever useful, not only to the student, as a " systematic code of probable errors," but to the practitioner as a memorial refresher of the difficulties which he has to encounter in the formation of a definite, well-grounded, and explicable opinion upon diseases of the thoracic viscera.

At the time when the paper on " Fatty Degeneration of the Liver " was written, a prevailing notion existed that it occurred only in association with phthisis. The lapse of thirty years since it was written has not diminished its interest; for the true pathology of the disease and its relation to many other morbid conditions are not yet fully understood. The purport of the paper was to show that fatty liver was not a mere accident attendant upon *phthisical or scrofulous* disease, but that it should be regarded as a more positive

than I: I saw the only child (a daughter) of one of the first noblemen of this country several times, and from the absence of physical signs, in spite of the dry cough, I assured him that there was no disease of the lung; an inflammation of this kind set in, and she died in six weeks. I have never been forgiven." Our patient died ten days after this interview. T. M. D.

affection; either as a primary or substantive malady, or indicative of a wide-spread morbid process affecting the whole body. Addison had observed that in many forms of cachexia a fatty liver was the most important, or, at least, the most obvious morbid alteration found in the body after death: he noticed also that it was associated with a fatty change elsewhere, and that patients suffering from it exhibited a peculiar anæmic aspect, and possessed a *remarkably white, soft, and smooth skin*. It was to the recognition of fatty liver by these signs that he more particularly drew the attention of his readers.

We may state that, at the present time, when a large fatty liver is found after death as the most striking morbid condition present, it cannot always be satisfactorily determined how far it is indicative simply of a fatty decay of all the tissues, or how far it is to be regarded as the essential disease.

This short paper shows that Addison was fully alive to the importance of the subject.

To the minds of those who can look back upon the ill-defined pathology of lung disease which existed thirty-five years since, notwithstanding the memorable, exact, and incomparable teachings of the great Laennec *up to a certain point*, and the paper of Dr. Alison read before the Medico-Chirurgical Society of Edinburgh, 1822* (of which there is no evidence that

* This great, thoughtful, and philosophic physician then read an elaborate paper on the "Pathology of Scrofulous Diseases." It included, of course, that of phthisis, and he enlisted the testimony of the accurate

Addison had any cognisance). We believe that a careful study of these five collective papers will convey an accurate impression of the definite, staunch, and truth-seeking mind, which, with the aid of an indomitable will, enabled Addison to unravel what must have appeared to him *then* the tangled problem of slow lung disorganization.

To those less advanced in life, who have imbibed from their instructors much of what Addison taught, without a knowledge of its derivation, it will convey the useful lesson that there is one *only* way of overcoming the difficulties which beset practical medicine; the way adopted by Addison. *Nothing short of a laborious investigation of clinical facts and the pathological corroboration of them could satisfy his accuracy and his rigid causality.*

Dr. Abercrombie as to the variety of morbid structure found in the lung, after death from phthsis, at different periods of life; but neither Alison nor Abercrombie appear to have formed any accurate appreciation of the textural changes which had occurred in the lung, nor the relation of those changes to the actual existence or non-existence of tubercle.

OBSERVATIONS

ON THE

ANATOMY OF THE LUNGS.

(*Read before the Royal Medical and Chirurgical Society*, 1840.)

MANY others must have felt as I do, that it requires some nerve to make a demand upon the time and attention of such a society as this; they must have been conscious, as I am, that however interesting or apparently important the intended communication may have been in their own estimation, it may, nevertheless, be deemed by the society at large altogether unworthy the character and experience of the majority of its members.

On the present occasion I hardly know whether such apprehensions ought to be diminished or enhanced by the brevity and incompleteness of this communication, its brevity and incompleteness being almost a necessary consequence of its subject, forming a mere part or portion of the results of a much more extended inquiry into the healthy and morbid anatomy of the lungs—an inquiry by far too extensive and of too much practical importance to be quickly accomplished, or hastily presented to the notice of the profession.

In the presence of any body of professional men in any part of the world, but especially in our own country, it would be in exceedingly bad taste to insist upon the importance of all and everything appertaining to either the healthy or morbid anatomy of the lungs; nor would it be less superfluous to add anything to what has been so repeatedly urged in proof of the necessity of acquiring an intimate knowledge of the former in order clearly to understand and satisfactorily to demonstrate the latter; it is not, however, by any means so certain that the converse of this familiar proposition has met with its full and legitimate share of attention; it is by no means so apparent that morbid anatomy has been sufficiently rendered available to explain or demonstrate the function or structure of obscure, intricate, and complicated parts of the body in a state of health.

For my own part, if I may be pardoned the egotism, I would venture to express my belief that very much assistance indeed may be derived from this source; whilst, on the other hand, the increase of knowledge thus acquired respecting healthy anatomy prepares us for still further advances in the investigation of changes induced by disease.

Proceeding on this principle of rendering healthy and morbid anatomy mutually subservient to the elucidation of each other, I am not without a hope that I have succeeded in illustrating, if not in demonstrating, certain points of great interest as regards the lungs, some of which have already been pretty generally established; whilst others are familiar as mere matters of opinion, or are perhaps altogether novel.

The results of my investigations, for example, seem to prove almost beyond dispute, first, that the aerial cellular tissue of the lungs is made up of well-defined, rounded, or oval lobules, united to each other by interlobular cellular membrane, each lobule constituting a sort of distinct lung in miniature, having its own separate artery and vein. Secondly, that these lobules do not communicate directly with each other. Thirdly, that they do not, as Reissessen and others have supposed, consist of the globular extremities of as many bronchial tubes, but, on the contrary, as my friend Dr. Hodgkin has suggested, are made up of a collection of cells, in which, by a common opening, a minute filiform bronchial tube abruptly terminates. Fourthly, that the pulmonary artery accompanies the bronchi, branch for branch, to the minutest divisions of the latter. Fifthly, that pneumonia consists essentially in inflammation of the aerial cells. Sixthly, that pneumonia and inflammatory tubercle are identical. Seventhly, that acute pneumonia in moderately good constitutions scarcely ever leads to the formation of an abscess, unless deposit previously existed; but that when it occurs in cachectic or broken-down constitutions, or supervenes in the process of chronic or organic diseases, it occasionally causes one or more distinct and separate lobules to soften down into an ill-conditioned abscess. Eighthly, that ordinary tubercles present the same varieties in the lungs as they do in serous membranes. Ninthly, that emphysema of the lungs consists chiefly of mere dilatation of the cells, but in part also sometimes of more or less extensive laceration of them. And, lastly, that the circumscribed gangrene of Laennec is commonly, if not uniformly, a mere effect or advanced stage of pulmonary apoplexy. It is not, however, to any of these matters that I presume to solicit attention at present, my object on this occasion being merely to point out a mode of distribution of the pulmonary vein, which, so far as I know, has not been noticed by any preceding anatomist or pathologist, and which, on that account, I have thought would prove not unacceptable to the members of this society.

The uncertainty and discrepancy which at present prevail respecting the origin and course of the pulmonary vein may be gathered from the descriptions given of it in the works of some of the most distinguished modern authorities, the quotations of a few of which may not, perhaps, be altogether superfluous or misplaced here.

Cloquet, in his large work, 'Anatomie de l'Homme,' sec. 509, observes :—"Les dernières ramifications de l'artère pulmonaire donnent naissance aux radicules des veines du même nom. Ces veines qui augmentent successivement de volume, *marchant à côté et au-dessous des artères correspondantes*," &c. &c.

Meckel, in his 'Manuel d'Anatomie,' translated and amended by Jourdan and Breschet, tom. iii, p. 518, in speaking of the lungs, says, "Dans l'interieur de l'organe, les veines pulmonaires accompagnent les ramifications bronchiques de plus près que les artères."

Adelon, again, in his 'Physiologie de l'Homme,' seconde edition, tom. iii, p. 147, on the subject of the pulmonary veins, observes, "Ces veines commencent par des radicules qui sont aussi inapercevables, et par consequent aussi peu connus que les dernières ramifications des bronches et de l'artère pulmonaire : *disséminés dans le parenchyme du poumon*, peut-être continus aux ramifications de l'artère pulmonaire. *Situés probablement aux mêmes lieux où aboutissent ces ramifications et celles des bronches et où* se fait la respiration," &c. &c.

Ollivier, in the article Poumon of the 'Dictionnaire de Médecine,' observes, " Que les troues des artères et pénétrent dans la poumon et en sortent par le même porrit *et que les ramifications des veines pulmonaires sont plus voisines des canaux bronchiques que celles des artères pulmonaires.*"

Bichat,[1] on the same subject, speaks thus :—" Les veines pulmonaires nées du système capillaire de ces organes, *suivent une*

[1] 'Anatomie Descriptive,' tom. iv, p. 65.

direction analogue à celle des divisions artérielles. Voisines, des leur origine, des ramuscules aériens, elles n' abandonnent jamais ces conduits dans leur trajet. Elles se réunissent successivement en rameaux plus volumineux que l'on trouve toujours appliqués sur des rameaux bronchiques d'un volume proportionné. Ordinairement la veine est inferieure à la bronche tandis que l'artère lui est superieure; c'est du moins ce que l'on trouve tant que l'œil peut suivre ces vaisseaux."

In none of the accounts given by these high authorities do I discover the slightest allusion to the mode of distribution of this important vessel, which my own dissections appear so clearly and unequivocally to demonstrate; but, on the contrary, find them to be totally at variance with it.

In order to accomplish this demonstration the pulmonary artery was injected with size, coloured red, whilst the vein was injected with the same material coloured yellow. The lung was then laid aside, and kept moistened in a cool place for several days, with the view of softening, by approaching decomposition, the connecting cellular membrane distributed throughout the lungs.

In this way the common cellular membrane beneath the pleura became so lacerable that the pleura itself was stripped off without much difficulty, and without inflicting any breach whatever in the aerial cellular structure of the lung which it had covered.

The lung, thus divested of its pleura, presents to the eye, more or less distinctly, lines on its surface which indicate the situation of what may be called the *pulmonary fissures*—a term more correctly applicable than that of interlobular, inasmuch as by the term interlobular is usually understood a something situated between either the longer lobes or smaller lobules; whereas by the term *pulmonary fissures* is meant certain spaces occupied by common cellular membrane, and which descend from the surface towards the interior, but without penetrating the aerial cellular tissue of the lung, thereby dividing more or less deeply the surface of the organ into a number of insular portions, some of which may comprise a great number of lobules. Guided by the linear indications on the surface of the now naked lung, we can in general, with the aid of a pair of points let into handles, or a pair of fine scissors, and without much difficulty, succeed in laying open and exposing the pulmonary fissures, at the bottom of which, merely surrounded by a loose cellular membrane, and resting on the unbroken aerial pulmonary tissue, we discover a vessel; that vessel is the pulmonary vein, alone, and unaccompanied by any artery whatever.

This vessel may be distinctly traced from larger to smaller trunks towards its source, until we reach the common cellular membrane between the ultimate lobules, from the exterior of which the vein appears to originate; whilst on the other hand, by continuing the mechanical operation towards the root of the lungs, we with almost equal facility trace the vessels still lying at the bottom of the pulmonary fissures, and becoming gradually larger and larger by the addition of branches, which proceed into the pulmonary fissure, and are derived either from the neighbouring smaller pulmonary fissures, or from the uniting cellular membrane between the ultimate lobules themselves, until at length it joins the large trunks at the root of the lungs, to form the great pulmonary veins. A small artery is not unfrequently observed running across the pulmonary fissures, from a portion of lung on one side to a portion of lung on the other; and in one instance I have found an exceedingly narrow strip of healthy lung passing like a bridge across the fissure on the very surface of the lung.

Thus, then, the human lung may be said to be made up essentially of a vast expanse of membrane, the interior of which during the whole of extra-uterine life is unceasingly exposed to the influence of atmospheric air, and upon the surface or in the substance of which are spread out the capillary ramifications of the pulmonary artery; these arterial capillaries passing from thence to the exterior of the membrane to form the pulmonary vein, which throughout its whole course is found to be situated on the exterior of the aerial cellular structure of the organs. It is unwise to be too sanguine, yet I cannot help indulging a hope that with a knowledge of the striking and distinct distribution of the pulmonary vein, we shall be more successful in our investigations into some of the most interesting and important diseases of the lungs; that it will be the means of throwing additional light on the origin and progress of that fatal scourge, phthisis pulmonalis; and that it will enable us, almost without a doubt or difficulty, to set at rest the long agitated questions respecting the origin and seat of pulmonary apoplexy, and more especially of what has been called œdema pulmonum, or dropsy of the lungs.

In these investigations I am at present engaged, and may take the liberty of adding that for the future I hope to have the co-operation of my friend and colleague, Mr. Hilton, demonstrator of morbid anatomy at Guy's Hospital, to whom I am already much indebted, and from whose extensive acquirements both as an anatomist and pathologist I anticipate much assistance and instruction.

OBSERVATIONS

ON THE

DIAGNOSIS OF PNEUMONIA.

Read before the Guy's Physical Society, 1837.

ANY attempt at a further elucidation of pneumonia, after the splendid performances of Laennec, may probably appear presumptuous; and especially so, when made by one who acknowledges himself indebted for almost all that he knows of thoracic diseases to that truly great man, at once the most distinguished and most successful cultivator of medical science that ever adorned the profession. When, however, it is recollected how vast and barren was the field of his inquiries when he commenced his brilliant career—and when our former ignorance is compared with the knowledge that resulted from his unprecedented discoveries—our astonishment is, not that he should have left something undone, but that he should have done so much. It is with the most profound deference and respect for his memory, therefore, that I venture to add this tributary mite to the riches of one of his favourite essays. I cannot but feel, also, that some apology is due to the profession, for presuming to direct attention to a subject with which the works of Laennec must already have made them familiar, and particularly to those who have so far resisted the influence of prejudice as to have made themselves conversant with the use of the stethoscope. My apology is that the very familiarity of the subject appears to have lulled medical men in general, and even the stethoscopist, into a too passive confidence in what is already known: and has probably proved a check to that correction and improvement which Laennec himself was at all times so eager to accomplish.

The main object of this brief communication is, to make some addition, however trifling, to the ordinary means of diagnosis; since experience has forced upon me the conviction, that there are few acute diseases more frequently mistaken or overlooked than pneumonia, to the detriment of the patient, and the no small embarrassment of the practitioner.

In order to make myself understood, I may perhaps be permitted to take a very slight survey of the pathology, signs, and symptoms of the disease; merely observing at the outset, that, in doing so, I shall adhere as closely as possible to the purely practical tenor of our Reports; indulging in theory no more than is unavoidable, in arranging and reasoning upon facts derived from the sick chamber and the dissecting-room. To the facts, or supposed facts, alone, do I attach any importance. The use of these facts must be left to the judgment and discernment of the reader.

In pneumonia, the inflammation is manifestly seated in or around the air-cells, or in both situations. It is, perhaps, of little importance whether we conclude it to be seated primarily and essentially in the one or in the other of these structures; although, for my own part, I entertain no doubt whatever of its being primarily and essentially seated in the interior of the cells themselves—a belief drawn from the successive local changes observed to take place as the disease advances. In the first stage of the disorder, we find the cells red, and filled with a serous-looking and sometimes bloody fluid, rendering the lungs more heavy, dense, and œdematous, whilst they still retain their tenacity. At a more advanced period, or second stage, the cells are found filled up with red solid matter, which appears to consist of the thickened parietes of the cells themselves; for if the lung be torn, and the torn surface examined with a magnifying-glass, it seems to be made up of innumerable minute red grains, just such as one might conceive to result from a filling-up of the cells in the manner supposed. At this period, the serous-looking fluid has disappeared, the lung is comparatively dry, and the tenacity of the solidified part is so far diminished that it may be readily broken down by forcing the finger into it: this is what has been called red hepatization. At a later period, and sometimes apparently without having been preceded by the red granules, the solidified lung presents a grey appearance, an albuminous matter seems to occupy the place of the granules, or rather their centres, constituting the grey hepatization. This albuminous matter is sometimes firm and fixed, at other times it is less plastic, and occasionally, especially in bad constitutions, takes on a more decidedly purulent aspect, and may be squeezed out by pressure; or, as the cohesion of the pulmonary tissue is often, under such circumstances, very much diminished, the slightest pressure of the finger causes it to break down into a semi-fluid mass, resembling an abscess.

It is not necessary to be more minute in describing the pathological changes which take place in the progress of pneumonia; it is sufficient to remember, that, in the first stage, the cells contain air and a serous-looking and sometimes bloody fluid, as shown by the peculiar crackling sound, and escape of the fluid on squeezing a cut surface; that, in the second stage, the cells are solidified comparatively dry, and, sooner or later, have poured into them an albuminous matter, either solid and fixed, or, more rarely, a matter approaching the character of pus. The stethoscopic signs indicative of these respective changes are such as might be expected, and are easily understood. Whilst the cells contain air and serous fluid, there is little or no dulness of sound on percussion, but during respiration, we hear the crepitating rattle—a rattle which undoubtedly depends upon the presence of air and fluid in the cells, for it is observed in cases of œdema of the lungs, and in some instances of pituitous catarrh, as well as in the first stage of pneumonia. When the cells are solidified and admit no air we have dulness of sound on percussion, bronchophony, and bronchial respiration, at least when the consolidation is considerable, and seated near the surface. Such are the stethoscopic signs of simple pneumonia, they are quite characteristic, and are pretty uniformly present, except under very peculiar circumstances.

If an opportunity present itself of examining the body, when a lung consolidated by pneumonia is retrograding towards a recovery of its normal state, we commonly find the cut surface of the portion previously hepatized of a pale or pinkish hue; or we find it presenting a mixture of pale, pink, and grey: it is still more friable and lacerable than natural; and the cells are again more or less loaded with serous-looking fluid, rendered frothy by squeezing the lung, in consequence of the presence of a considerable number of air-bubbles. It would also appear that the further changes consist in the absorption of the effused fluids, a gradual increase of the tenacity of the pulmonary tissue, and a more or less complete restoration of the normal state. In some instances, however, when the albuminous matter thrown out is of the more plastic or organisable kind, it fails to be entirely absorbed, and part of it permanently remains. Under these circumstances, we find it, at an after period, either in small, detached, and more or less rounded masses, or more extensively and more irregularly diffused through the pulmonary tissue. When distributed in small insulated portions, I believe it to constitute one of the forms of albuminous deposit, indiscriminately called tubercles: whereas, when more extensively and irregularly diffused, it has, in like manner, been regarded as a form of tubercular infiltration. The history, however, of the patient's case, in many instances, as well as the local appearances themselves, lead me to the conclusion that they are merely the result of a previous attack of pneumonia. We often learn on inquiry that at some former period, perhaps years before, the patient had had an attack of inflammation within the chest; whilst, if he die of some other disease, we almost uniformly discover on dissection unequivocal evidence of antecedent inflammation. The evidence consists in thickening and adhesions of the pleuræ, especially in the neighbourhood of the appearances in question, together with induration and puckering of the pulmonary tissue immediately surrounding each albuminous deposit: or, when the deposit is irregular and extensive, we often have an actual deformity and puckering of the pleura above the infiltrated parts. This view of the origin of these albuminous deposits will probably serve in some measure to explain why they are much less uniformly found in the apices of the lungs than ordinary tubercles.

It has been observed that these deposits may remain passive for an unlimited period, and without undergoing any very appreciable change, except perhaps a conversion of some of them into calcareous or chalky masses, especially when deposited in the upper lobe of the lung: it would nevertheless appear that the vital influence by which they are maintained in their integrity is so extremely slender, that if inflammation happen to be set up around them by any accidental cause, and especially if the vital powers of the patient have been greatly impaired, that influence is so far exhausted, that they lose their cohesion and soften; the softening commonly first taking place in those portions most remote from the more highly organized living structures; they soften in the centre; the softening proceeds outwards, and in the end causes the formation of a vomica, and so produces one of the modifications of phthisis pulmonalis. Such, at least, are the conclusions to which repeated observation of the living and dissection of the dead have led me in regard of this part of the subject.

Having premised these very superficial remarks, I shall now proceed to the repeated functional signs or symptoms of pneumonia;

for it is to the unsteadiness and fallaciousness of these that errors in diagnosis are chiefly attributable; and, consequently, it is to them more particularly that I am desirous of directing attention.

The characteristic symptoms of pneumonia enumerated by Laennec are, an *obtuse and deep-seated pain in the chest, dyspnœa, hurried respiration, cough,* and *peculiar expectoration;* but, in reference to these, he tells us that each of them individually may occasionally be absent, and, indeed, that they may all be absent in the same case. Now, were it quite correct to assume that the character of pneumonia is that which is expressed by the above symptoms, that the reputed deviations and exceptions alluded to by Laennec are only of very rare occurrence, and that obscurity happened only in the pneumonia of old people, and in cases complicated with other diseases, there might probably be some excuse for resting satisfied with the present position of the subject; but if it be as true, as I am convinced it is, that these reputed deviations and exceptions, regarded as obscure, are of extremely frequent occurrence, that they are met with at every period of life, and in every variety of constitution; and that they are very far indeed from being limited to old persons, and to what have been called complicated cases; I hope to be pardoned if I make an attempt in some degree to unravel the difficulty, and place the subject, if not in a more correct, at least in a more safe and practical point of view.

I have been led to the conclusion that cases of pneumonia, characterised by obtuse and deep-seated pain, dyspnœa, hurried respiration, cough, and peculiar expectoration, are, in truth, themselves the exceptions in a pathological sense; and that, although most frequently met with in practice, they are, in fact, cases of complication. It may be said if such cases of complication be those most commonly encountered in practice, why interpose a mere pathological subtlety to disturb the practical rule? To this I oppose my belief that it is an adherence to such a general character of pneumonia that has led, and is constantly leading, to an oversight—to a neglect of the disease, when it occurs in what I am disposed to regard as its more *simple form;* and as cases partaking more or less of this simple form of pneumonia are of frequent occurrence, I am willing to persuade myself that what follows may have the effect of diminishing the liability to the errors alluded to.

In *simple pneumonia,* after chilliness, shivering, feebleness, and depression, the patient experiences, for the most part, strongly-marked symptoms of febrile reaction, giddiness, confusion, and sometimes intense pain in the head; occasionally delirium, especially towards night; *the skin acquires a pungent heat,* generally accompanied by dryness, more rarely by moisture; the pulse is full and strong, perhaps labouring and sluggish; the face is usually more or less suffused with a livid flush, accompanied by an expression of distress; the tongue is foul, its substance is more injected than in ordinary phlegmasiæ, and in a short time it manifests a tendency to become dry and brownish; the respiration is somewhat hurried, but *there is seldom any very obvious cough or expectoration, and sometimes none at all;* in short, the whole assemblage of symptoms bears a most striking resemblance to those of a severe attack of common, continued fever of the typhoid type, for which it is so repeatedly mistaken. If this form of the disease occur in moderately good constitutions, and is overlooked, especially if stimulants be administered on the supposition of its being a severe case of typhoid fever, it very commonly happens that the general prostration increases, the delirium or oppression of the brain is aggravated, the tongue gets dry and black, and the teeth covered with sordes; the breathing becomes more hurried, occasionally with a frequent slight hacking cough, and now and then a little bloody expectoration; the pulse gets flaccid, frequent, and feeble; and at length the patient dies.

Notwithstanding its close resemblance to a severe attack of continued fever—a resemblance so great that even the stethoscopist is occasionally thrown off his guard—attentive observation will, in most cases, enable us to recognise the difference. The attack in general is more abrupt, and often follows some manifest exposure to cold or wet. The countenance, though congested and somewhat distressed, has not the dejection and stupidity so remarkable in fever; it displays more intelligence; and, although confused and perhaps slightly delirious, the patient, on being roused, commonly evinces a clearness and vigour of intellect not found in fever. The condition of the tongue also furnishes a valuable diagnostic sign. We know that, at the onset of the fever, the contrast between the vividly injected tongue and its white or grey fur is very striking; it is, in general, much less so in pneumonia. In the latter, if I may be allowed the expression, it is more the tongue of a phlegmasia; the hurry of respiration in pneumonia is often not more than we commonly perceive amid the general distress of fever; and I repeat that neither cough nor expectoration is necessarily present in a very appreciable degree. But of all the symptoms of pneumonia, the most constant and conclusive, in a diagnostic point of view, is *a pungent heat of the surface*. By this symptom alone the first stage of pneumonia may in most instances be readily recognised. By this symptom alone I have repeatedly pronounced the existence of pneumonia before asking a single question, or making the slightest stethoscopic examination of the chest. The presence of this symptom has scarcely ever yet deceived me, even in the most complicated forms of inflammation within the chest. I by no means contend that it is necessarily present at some period of every case, although I do not know to the contrary; but I feel justified in affirming that when inflammation is confined to the chest, however varied may be the tissues involved in the inflammatory process, provided this symptom be present, pneumonia may be confidently pronounced to form a part in nineteen cases out of twenty, and I believe in a much larger proportion.

A similar pungent heat of the surface is now and then observed in certain forms of renal dropsy; more frequently in continued fever, especially in children; and still more commonly in the eruptive fevers of the exanthemata and erysipelas; and as such cases may supervene upon already existing disease within the chest, the fact ought to be carefully remembered, lest the most valuable diagnostic sign should rather mislead than assist us. It is in original inflammation within the chest that it proves so constant and conclusive a sign of pneumonia; but on every occasion when present it ought to lead to a most careful scrutiny by means of the stethoscope.

I am unwilling to swell this communication by a detailed recital of individual cases: but were it otherwise it would be easy to introduce a very great variety of instances in which simple pneumonia has been mistaken for common fever of a typhoid type. I have repeatedly witnessed it in children, the first suspicion of it having generally been suggested to me by recognising, on applying the hand to the surface, the peculiar pungent heat already noticed. I not long ago had an example in a young woman who was supposed to be labouring under a severe attack of bilious fever; so called because pneumonia of the right lung was accompanied, as is not unfrequently the case, by a sallowness or almost jaundiced aspect of the patient's countenance. I have a very similar case in Miriam's Ward at this time, also occurring in a young female. In elderly persons it is so common that when a case of typhus is represented to have occurred in any individual above fifty years of age, without evidence of the existence of the disease in other branches of the family, I confess that I consider it at all times an equal chance that it is, in reality, a case of pneumonia. An instance of this kind I saw very recently: the person was upwards of sixty, but of a hale constitution, and presented most of the ordinary signs of continued fever, whilst the pulmonic symptoms were so slight as never to have attracted the least attention. This brief representation may probably suffice to fix attention upon the likelihood of the presence of pneumonia in cases of supposed continued fever.

The more simple form of pneumonia not unfrequently assumes another appearance, which has occasionally led to a belief that the brain was the seat of the disorder: the original affection of the lungs being so obscure as to be entirely overlooked. I have, within a short period, seen two cases of acute pneumonia in vigorous adults, in which, at the commencement, and for some days, the disturbance of the brain was such, that remedies were applied exclusively for the relief of that organ. In both instances the inflammation was very intense, and was, at a later period, attended with cough, expectoration, and other signs commonly regarded as characteristic of pneumonia.

Some time ago I was requested to see an elderly man, who appeared to be labouring under obscure symptoms of mental aberration, and was supposed to have become insane. He looked pale, his countenance was somewhat anxious, his tongue was loaded, slightly brown, and disposed to become dry, he was occasionally incoherent, and wandered about the ward in a wild and unaccountable manner, but had neither cough nor expectoration sufficient to attract any particular attention. On examination I found him labouring under pneumonia already advanced to hepatization, He recovered. A similar case is now under treatment in the hospital.

In infants and very young children such cases are by no means rare, and simulate hydrocephalus. In one instance, where hepatization had taken place, the most prominent symptom was convulsions, for which various applications had been made to the head.

Such are some of the affections of the brain, to which pneumonia not unfrequently gives rise—secondary affections, calculated to mislead the most wary; and such as must inevitably distract the attention, and perplex the judgment of those who do not habitually have recourse to the stethoscope.

DIAGNOSIS OF PNEUMONIA. 15

If the representations I have made be correct they certainly lead to an inference, that even acute disease does not, when confined to the air-cells, necessarily give rise either to cough or expectoration—symptoms, perhaps, too much relied upon in recognising, or even suspecting, affections of the lungs.

Without arguing the question, whether it be possible to expectorate a thin watery fluid, which must necessarily gravitate in the cells of the lungs, I may venture to state that I entertain a very strong suspicion that the cough and expectoration so commonly observed in pneumonia depend altogether upon the accidental implication of the bronchial tubes, and that, without a doubt, the degree of these symptoms depends upon the degree of that implication. Certain it is that the most intense pneumonia may exist, even in hale constitutions, with cough and expectoration so slight as to pass unnoticed; and it is not difficult to suppose that, when so slight, they may depend rather upon mere sympathetic irritation of the minute bronchial tubes in the immediate neighbourhood of the inflamed tissue, than upon any considerable degree of actual inflammation set up in them. It is true that, on dissection, we very commonly find the mucous membrane of the smaller tubes reddened; but whether from inflammation or not, is by no means so easily determined. I am disposed to think that, in simple pneumonia, the small tubes are either not at all inflamed, or only inflamed in a very slight degree, and that, when more decidedly involved, their inflamed state gives rise to the cough and peculiar viscid expectoration described as characteristic of pneumonia in general. This complication is indisputably more frequently present than absent; a circumstance little calculated to excite surprise, and one probably sufficient to account for the symptoms which attend the complication, having usually been described as those essential to, and characteristic of, pneumonia.

When cough and expectoration are as well marked as they are commonly described to be, they cannot fail to attract the attention of every one, and all difficulty of diagnosis ceases. The same may be said of those cases of pneumonia in which we have the mucous membrane of the bronchial tubes involved to such an extent, that, by universal consent, the disease is said to be complicated with bronchitis, and in which we have the expectorated mucus, though considerable in quantity, more or less tinged of a brownish or saffron colour. It does not, however, necessarily follow that, when pneumonia is present, the mucus of the accompanying bronchitis shall be tinged brown; on the contrary, the discoloration is often in such cases, altogether absent, its presence and degree depending upon the quantity of blood which happens to be diffused; exactly in the same manner as the ordinary *viscid sputa* of pneumonia may be colourless, or may be of a gamboge yellow, light green, or of a rusty or red colour, according to the same accidental circumstance. Of course, in these bronchial complications, we have, superadded to the stethoscopic signs already mentioned, a mucous rattle, which, when hepatization takes place, is rendered much more distinct, in consequence of the consolidated lung being a better conductor of sound.

In concluding this slender contribution to diagnosis, I shall merely observe further, that unless complicated with pleurisy, pain of any sort is rarely complained of by a patient affected with simple pneumonia, in whatever position he may be placed. When, however, the bronchial complication is such as to produce severe cough, he not unfrequently experiences a burning or tearing pain, or rather soreness, more or less diffused through the affected parts—a symptom probably resulting from the violence inflicted upon the inflamed tissue during the repeated fits of coughing.

OBSERVATIONS

ON

PNEUMONIA AND ITS CONSEQUENCES.

(*Read before the Guy's Physical Society*, 1843.)

My object in offering this communication to the Society is, to direct attention to a few points connected with the pathology, diagnosis, varieties, and effects of pneumonia. The general history and ordinary details appertaining to the disease are, I believe, sufficiently well known; and might be deemed misplaced, or might even appear impertinent, when addressed to a society which numbers amongst its members those who have successfully directed much of their attention to the subject. The disorder, nevertheless, is one in the study of which I have long felt a deep interest; and the result of my inquiries and investigations is a conviction that much still remains to be done for the satisfactory elucidation of those points to which I have alluded. However unsuccessful I may be in my attempt to add a trifle to the general stock, it is gratifying to hope that by exciting discussion and thereby concentrating the practical knowledge of the Society, much useful information may be elicited.

As I have already stated, I shall content myself with a very brief notice of those parts only of my subject which seem more particularly to require or admit of further elucidation: and in compliance with such design, I shall first direct attention to what has been termed *simple pneumonia*, by inquiring whether that which has been so called be in reality the simplest form of the disease, and whether the general impression that it is so has not led to many serious and even fatal errors.

Every one knows that the symptoms said to characterise *simple pneumonia* are phlegmasial fever, with dyspnœa, pain or uneasiness in some part of the chest, cough, and peculiar expectoration. It is true that Laennec and others have dwelt upon the variableness in degree of each of these symptoms, and have carefully pointed out a certain class of cases in which they do not appear, and which on that account have been designated *latent pneumonia*; cases occurring in the aged, the cachectic, in the intemperate, and towards the fatal termination of various diseases, characterised in general by much typhoid prostration, and thence occasionally called *typhoid pneumonia*. An impression has thus been created, and is still very generally prevalent, that whenever pneumonia occurs in good constitutions, and especially if the patient be young, the characteristic symptoms enumerated ought to be present. This impression is at variance with my own experience; and I am led to the conclusion that the simple pneumonia of Laennec and others is not the simplest form of the complaint, but a complication—a broncho-pneumonia, and that a truly simple pneumonia is not very unfrequently met with in young persons, and in good constitutions, unattended by either cough, expectoration, or pain, or, at least, such a degree of either as to attract particular attention; but more of this when we come to the diagnosis.

Seat of Pneumonia.—There are probably some present who remember the time and occasion when, in this Society, and in opposition to all existing authorities, I ventured to call in question the long cherished notion that pneumonia had its seat in a supposed parenchyma of the lungs, and that the products of pneumonic inflammation were poured into that parenchyma. Since that time I have had the satisfaction of witnessing a gradual, but comparatively rapid, renunciation of the latter views, and the adoption of those advanced for discussion in this Society so many years ago; viz., that pneumonia has its original and essential seat in the air-cells of the lungs, and that the ordinary pneumonic deposits are poured into these cells. It is nevertheless true that some of our most recent authorities are opposed to this opinion, and maintain that the pneumonic deposits are poured into an interstitial tissue, a conclusion which I find myself unable to reconcile with either the healthy or the morbid anatomy of the lungs. I entirely fail to discover any structure to which the terms *interstitial* and *parenchyma* can be fairly applied. Accompanied by a corresponding branch of the pulmonary artery, I trace a filiform bronchial tube to a lobule or bunch of cells, in which it abruptly terminates; the blood distributed over these cells being received by the pulmonary veins, which pass exteriorly to the air-cells, in a loose and very distinct interlobular cellular tissue.

But, notwithstanding the most careful investigation, aided by injections into the tubes and cells, and by the use of magnifying-glasses, I must confess myself unable to arrive at any positive conclusion, either as regards the elementary tissues which compose the air-cells of the lungs, or the exact construction and arrangement of the cells themselves. These are questions which there is reason to hope will ere long receive a satisfactory solution from those who have already distinguished themselves as successful cultivators of microscopic anatomy. In the mean time, without venturing to decide whether the innumerable, minute, irregular, and manifestly elastic air-cells constituting an individual lobule, partake more of the character of areolar tissue or of a serous membrane, I am fully persuaded that *pathologically* they present none of the attributes of a mucous membrane, as Reissessen and others would lead us to believe; and that, so far as the changes induced by inflammation and other diseased conditions are entitled to guide us, they must be ranked with one or other of the first-named tissues. Neither do these changes enable us to recognise any of that intercellular tissue which has been confidently asserted to be the seat of some of the most important diseases of the lungs, and into which, accordingly, the morbid deposits resulting from inflammation and dropsy have been supposed to be effused. So far as my observation has extended, I should have as little hope of finding this interstitial material in the pulmonary lobule as in a mere bundle of common areolar or cellular tissue.

Effects of Pneumonia.—I may venture to affirm that no one who carefully traces the effects, either immediate or remote, of pneumonic inflammation can fail to be struck with their close resemblance to those of inflammation affecting a serous membrane or the common uniting cellular tissue of the body; and if the pathological changes induced in any tissue constitute a legitimate foundation for an opinion respecting the physiological character of that tissue, one is almost irresistibly led to the conclusion that the air-cells of the lungs are a mere modification of one or other of these tissues. But to proceed.

It is a plausible conclusion of one of our ablest writers on pneumonic inflammation that in the earliest stage of that morbid condition there is a preternatural dryness of the air-cells, from an arrest of their natural secretions, and that this stage is characterised by an excited state of the respiratory act, and by a murmur louder than natural. To this belief I am inclined to subscribe, having recently met with a case in this hospital, which from the presence of a premonitory symptom, to be mentioned more particularly by-and-by, I anticipated would proceed to pneumonia, in which pneumonia actually took place, and in which this excited state of the respiration and a loud but rough respiratory murmur in the lung about to be affected, were strongly marked. Further observation is, nevertheless, certainly still required fully and satisfactorily to establish this —Dr. Stokes's position; and since the onset of inflammation as positively checks the secretion of a mucous as it is affirmed to do that of a serous membrane, the position, even if correct, bears but little upon the question regarding the mucous or serous character of the air-cells. It is the effect which immediately succeeds to this, and which is that usually first recognised by physical signs, being the stage of engorgement of authors, that seems most strongly to countenance the belief in the serous character of the pulmonary air-cells. This effect consists in an effusion of serum into the cavities of the cells themselves; an effect certainly at variance with the opinion of Reissessen and others, that the air-cells are merely the blind extremities of as many bronchial tubes, and that, like the latter, they are lined by an ordinary mucous membrane. It is some years ago since I first collected this fluid, and on testing it by heat, found it to be highly coagulable. It is true that there are those who entertain a different opinion, and contend that the effusion, though serous, is not poured into the cells, but into an interstitial tissue. This interstitial tissue I have never been able to discover, whereas any one, with a very little care, may satisfy himself that the serous fluid is poured into the cells themselves; he may do this by forcing, by gentle pressure, the fluid from the cells through the truncated bronchial tubes of the incised lung, and subjecting it to examination. But as a proof of its not being in the cells, we are told that at this stage of pneumonia expectoration is often absent, or, if present, does not consist of serum; which it is supposed ought to be the case if it were contained in the cells. It remains to be proved, however, that simple serum, which must necessarily gravitate, can be expectorated at all under ordinary circumstances; inasmuch as in order to accomplish expectoration, the matter to be expectorated must admit behind it a sufficient quantity of air to force it into and up the tubes, and we know that in the very worst forms of œdema of the lungs, there is often very little cough, and seldom any expectoration at all, or, if any, not necessarily of a serous character. It is, indeed, contended that the serous effusion in œdema of the lungs is not poured into the cells, but into the supposed interstitial tissue. The disproof of this, however, is just as easy and as obvious as that of the interstitial deposition of serum in pneumonia; as I have over and over again satisfied myself; and I may add that Mr. Hilton and myself collected the fluid from the truncated bronchi with great care and with the least possible violence, and found it to be purely serum. In order, also, to prove that it was not lodged in the interlobular tissue we separated an individual lobule, and found that by puncturing it a fluid purely serous was readily collected. Thus, then, as in other serous structures, we find that one of the earliest effects of inflammation of the pulmonary cells is an effusion of serum into them ; and that when there is dropsy of other serous structures, œdema or dropsy also takes place in them; illustrations of which are perpetually met with in Bright's disease.

In this stage of pneumonia, together with a highly red and vascular condition of the cells, we find the cells already more substantial and dense to the feel than natural, although still crepitating and containing air. Sooner or later, however, the turgid parietes of the cells obey the ordinary laws of an inflamed part, they lose their natural cohesion, swell, and encroaching on the cavities of the cells, cause the absorption of the serum previously effused, and by thus occupying or filling up these cavities, occasion first a dryness and brittleness, and afterwards the complete consolidation which constitutes the red hepatization of authors—a consolidation, nevertheless, accompanied, as might be expected, by such loss of cohesion as causes the lung readily to break down under pressure. From the examinations I have made of lungs in this state, I am disposed to agree with those who reject the notion of the consolidation being dependent in any degree upon actual effusion of solid albuminous matter into the cells: it seems to result entirely from the great vascularity, softening, and tumefaction of the parietes of the cells themselves, and hence probably the rapidity with which this often takes place in typhoïd or cachectic pneumonia, in consequence of that loss of tone which uniformly causes the various tissues speedily to soften under the disorganizing influence of inflammation. Accordingly, if we attentively examine a lung in this condition with a moderately magnifying power, we find the granular appearance so much dwelt upon by writers, but without any evidence of actual deposit in the cells.

In order more clearly to comprehend the subsequent changes

which take place in an inflamed lung, it is necessary to bear in mind the disorganizing tendency which uniformly accompanies the morbid process of acute inflammation wherever situated. The changes produced by this morbid process we know to consist not only in a loss of cohesion and more or less tumefaction of the inflamed tissue, but in a remarkable disposition in that tissue to return to a state more or less resembling that which forms the original basis of all tissues, viz., an albuminous material. This re-conversion of a tissue into albumen I would express by the term *albuminization*. The healthy organizing process converts albumen into the various natural tissues, whereas the disorganizing process of inflammation tends to destroy this organization, and to cause the tissues to return to an albuminous state. We accordingly find that as acute inflammation proceeds, the parietes of the cells become more and more opaque—how far from a merely molecular change, and how far from superadded albuminous deposit, it is difficult to determine; the minute blood-vessels are no longer visible, the parietes of the cells are not only thickened, but become exceedingly softened; and with this extreme loss of cohesion and apparently diminished vascular turgescence, we have undeniable proof that the cells now admit of an albuminous matter being poured into their cavities. This constitutes the grey hepatization of authors.

When the inflammation occurs in good constitutions this albuminous matter may be more or less solid, so as not to allow of its being pressed out; although, by careful washing, it may sometimes be partially removed, so as to restore imperfectly the cellular appearance of the lung; but should the inflammation occur in a bad constitution, or take on an atonic form, the albuminous effusion is more fluid, presenting a more or less yellow or even dirty muddy purulent appearance, and may be readily forced out by gently pressing or by pricking the cells which contain it. That this puriform fluid is really contained in the cells themselves, and not in any interstitial tissue, may be further shown by making an incision into the lung, scraping off the matter from the divided surface, and then applying pressure; when the matter will be seen to issue from separate openings more or less apart from each other, and which are manifestly small bronchial tubes. These changes may, in any state of constitution, take place in a considerable extent of pulmonary tissue, rendering the whole of an opaque white or grey colour, and obliterating, or rather concealing, the lines of common cellular membrane which naturally separate one lobule from another, and which were, during the period of red hepatization, distinctly visible. In other cases, however, and especially in certain atonic forms of acute pneumonia, these changes are confined to individual lobules more or less remote from each other, the common cellular membrane appearing to form a distinct line or boundary to the inflammation; thus illustrating one of the purposes answered by the lung being divided and subdivided into lobes and lobules, that of preventing, to a certain degree, the extension of disease commencing in one lobule to that immediately adjacent.

It has long been a matter of remark that acute pneumonia rarely terminates in the formation of abscess; and it certainly is not a little remarkable that, notwithstanding the great loss of cohesion of the tissues—so great that the slightest violence after death causes them to break down into a semifluid amorphous mass—a real abscess should be so very rarely met with. In good constitutions, even under the most intense inflammation, such an occurrence is universally acknowledged to be extremely rare; whilst in the more atonic inflammations affecting bad constitutions, although abscesses are by no means so very unfrequent, they are for the most part exceedingly small, and irregularly scattered through the inflamed part. In some cases such is the want of vital power, that acute inflammation rapidly disorganizes the tissue, and either reduces it speedily to a puriform albuminous mass, or entirely robs it of its vital properties, so that it dies or sloughs before the softening and liquefaction are so complete as we find in abscess : in these cases of sloughy abscess the sloughing part sometimes presents a dark dingy aspect, exudes a turbid serous or sanious fluid, and even manifests an offensive or putrid odour. Gangrene, properly so called, is acknowledged to be still more rare; although, when gangrene of the lung does occur in its diffuse form, I believe it to be closely allied to the sloughing process just alluded to, and to result from a certain degree of inflammation occurring in bad constitutions; all the cases that I have witnessed having been in persons who had either been guilty of great intemperance, or had otherwise brought their systems into an extremely cachectic or atonic state.

I have, in these remarks, specified diffused gangrene in particular; because another form of gangrene of the lung is spoken of, of a more chronic kind, generally surrounded by condensed lung, and to which the name *circumscribed* gangrene has been applied. Most, if not all cases of the latter kind, however, I am disposed to suspect, have quite a different origin from the former, and are the result of previous pulmonary apoplexy; an illustration of which will be found in the accompanying drawings, taken from subjects examined at this hospital. Blood is poured into the aërial cellular tissue: this for a certain time remains in the form of a circumscribed solid red body, imbedded in the lung: it afterwards assumes the appearance of a mass of solid albuminous matter, the red particles or coloured matter having been removed. This albuminous mass may remain permanently, and be accidentally discovered when the individual has been cut off by some other disease. It would appear, however, from cases such as these from which the drawings were taken, that these masses occasionally destroy the circulation of the air-cells in which they are imbedded, act as foreign bodies, excite inflammation, and cause albuminous matter to be poured out around them, by which they become more or less isolated from the surrounding tissues, and consequently die; or, becoming putrid from their communication with a bronchial tube, give rise to all the ordinary characteristic symptoms of gangrene of the lung. It is in such cases that a portion of separated and dead lung is occasionally found as a loose nodule in the gangrenous cavity.

I have said that the changes produced by acute pneumonia may be limited to separate lobules, more or less remote from each other. This *lobular pneumonia,* as it has been called, although it may occur in good constitutions—as, for example, in pneumonic complications of whooping-cough—is, for the most part, observed in bad and cachectic habits of body, and especially towards the termination of various chronic diseases, after surgical operations, and in phlebitis. It is in such cases that the vital properties of the tissue inflamed are rapidly exhausted, so that remarkable softening or small sloughy abscesses speedily result, so rapidly, indeed, that, in the case of phlebitis, it has been somewhat unphilosophically, and certainly very incomprehensibly asserted, that the matter formed in the vein is mechanically conveyed to and deposited in the lung. The abscess in such cases is, indeed, sometimes of considerable size—perhaps that of a large hazle nut—very complete, and very rapidly formed; but that this merely results from the circumstances just pointed out is, I think, sufficiently attested by the fact, that in some instances we can distinctly trace, in different lobules, the progressive changes which usually precede a perfect abscess; namely, redness, dryness, condensation, friability, albuminous opacity, and, lastly, the softening down into a complete abscess.

There is another form of consolidation of the lung, which has been much dwelt upon by writers on this subject. It was first, I believe, particularly noticed by Laennec, who described it under the name of *carnification* of the lung, and was regarded by him as the result of pneumonic inflammation, modified by pleuritic effusion. Externally the lung presents a bluish tint, not unlike that of the spleen; it feels soft and flabby; and, when cut into, displays no granular, but rather a uniform texture, of a dark or brownish hue, pretty closely resembling raw flesh which has become dark-coloured from long keeping; it is totally destitute of air, does not crepitate, and instantly sinks in water. This, I repeat, has been considered as a consolidation resulting from pneumonia modified by pleuritic effusion; and has even been made the grounds of an argument, that pneumonic deposits do not take place into the air-cells; the deposit in the present case being, or supposed to be, seated in the interstitial tissue. I will confess that I never could understand these views, which are altogether at variance with what I understand by inflammation. I never could reconcile it to myself that an inflamed tissue could retain its normal, much less have an increased, tenacity; for the lung in this state is remarkably tough; whereas loss of cohesion is one of the most constant and inevitable consequences of inflammation in any tissue. I never could entertain a doubt that the whole of the appearances of such a lung as has been described resulted from pleuritic effusion having compressed the lung, forced out the air, and thus brought the sides of the cells into close contact. Accordingly I recently took a portion of lung in this state, and, having ascertained that it instantly sank in water, requested my friend Mr. Hilton to try to inflate it: he did so, and the result may be seen in the dried lung which I now present for inspection. In short, the lung is perfectly inflated; and the aërial cellular tissue is in its normal condition, with the exception of a slight redness, indicative of an unusual quantity of blood. That this may occasionally be blended with pneumonic inflammation and its consequences, I do not deny, though I do not know it; but I am strongly inclined to believe that the whole matter is one of those mistakes which, once made, continue to be handed down without correction, in consequence of a neglect to put it to the test of careful experiment.

I have met with instances of hepatization of the lung, more or less complete, presenting a uniform rather than a granular aspect when cut or torn; but these were altogether independent of pleuritic effusion. I have seen such cases after whooping-cough, and in hypertrophy of the heart. How much has mere hypertrophy of the aërial cellular tissue to do with such consolidation?

There is a peculiar state of lung frequently met with after death, the nature and origin of which are still involved in considerable obscurity. The condition alluded to is met with in the posterior or most depending part of the lower lobes of the lungs; and the first difficulty is to determine whether it existed before death, or is merely cadaveric, resulting from the position of the dead body. In most instances of this kind, the lung externally is of a dark colour, and when cut into presents an almost uniform soft, flabby, very humid, and readily-lacerable texture, in colour verging from that of red to that of black-currant jelly. If this take place after death, why is it found in one case and not in another? If it take place in any instance before death, what are the pathological conditions which determine the occurrence of this congestive softening, as it might then be called? Why I mention it in this place is because we very frequently find it mixed up with a certain degree of actual inflammation of the air-cells, as indicated by a marked difference of colour, and by the dryness and brittleness of the inflamed tissue; and now and then by an actual albuminous deposit, or purulent infiltration in certain points. In such cases did the congestion precede or give rise to the pneumonia, or was the apparent congestion purely cadaveric? In some instances, at least, the signs and symptoms enable us to predict one or both of the morbid appearances just described.

The *remote* consequences of pneumonia are scarcely less varied, and are certainly not less interesting than the immediate effects just described. The ordinary progressive changes which take place in a lung recovering from an attack of acute pneumonia are too familiar to require description; neither will I enter into any speculation as to that precise degree of pneumonic inflammation which does, and that which does not admit of perfect resolution—such resolution as shall restore permeability to the cells previously inflamed, without taking into the account any slight opacity or thickening insufficient to prevent their inflation.

The *permanent* effects produced in a lung by pneumonia depend, I believe, chiefly upon the state of the patient's constitution, and the character of the inflammation, and consequently upon the nature of the albuminous deposits, and the degree of organization of which that albuminous deposit is susceptible.[1] Of these permanent effects I have been led to distinguish three varieties: 1. *The uniform albuminous induration.* 2. *The granular induration.* 3. *The grey induration.*

When acute pneumonia occurs in good constitutions, the softened tissues appear occasionally to become so blended with or assimilated to the permanent albuminous deposit, that the whole is converted ultimately into a uniform homogenous, semi-transparent, or opaque and yellowish material, in which we discover not the slightest trace of either the aërial cellular structure, or of the common interlobular cellular membrane. Passing through this organized mass, we sometimes find vessels of considerable size carrying red blood, which, on making pressure, may be forced out of their truncated extremities; but as regards any proper vascularity of the new tissue itself, I am not at present in a position to speak with confidence. The material just described may either be diffused through a considerable portion of lung involving several lobules, or even the greater part of an entire lobe, which is sometimes the case; or, as more rarely happens, it may be limited to one, or a very few lobules only. This, the least frequent of the permanent pneumonic indurations of the lung, I would distinguish by the term, *uniform albuminous induration.*

When, on the other hand, the inflammation pours out a less organizable albumen, as often happens in strumous habits, the permanent change produced is very different. In such cases the interlobular cellular tissue often, though not necessarily, remains perfectly distinct; and a solid, pale, or yellowish and friable albuminous matter occupies the lobules; and this, apparently, without having assimilated the parietes of the cells, as in the former case. The cellular arrangement is still perceptible, both on the exterior and in the interior of the lobule, giving to the former something of the configuration of a raspberry, whilst, on cutting the lobule, we find the friable albuminous matter presenting the granular aspect

---

[1] Whenever the albuminous products of inflammation become consolidated and contracted, and remain permanently, I, perhaps with little propriety, use the terms "organized" and "organization." The state of the blood-vessels in and around these products may probably form a subject of the future communication.

produced by the filling up of still separate cells. Whether this change take place in a single lobule, in lobules at a distance from each other, or in a number of lobules as it were heaped together, so little affected is the interlobular cellular membrane in many instances that after keeping the lung a few days to soften it these lobules may be fairly turned out and examined; as was the case with the lobules from which the accompanying drawing was taken. This constitutes, then, what I would call the *granular induration*, resulting from a deposit the very character of which seems to indicate a less vigorous constitution, being less capable of organization, and resembling in its appearance ordinary tuberculous matter, from which latter circumstance it is sometimes called *inflammatory tubercle*.

The *grey induration* is made up of a mixture of dull or yellowish-white and black matter, in variable proportions; the tint being light or dark according to the greater or less quantity of one or the other of these; the lighter variety occasionally passing insensibly into the uniform albuminous induration, the darker, on the other hand, being of an iron-grey colour, or approaching, in some instances, nearly to black, the former again being in general of moderate density, the latter usually much firmer, and sometimes even as hard as cartilage. In this grey variety of pneumonic induration, the morbid change differs from the uniform albuminous induration, in the albuminization of the tissues being much less complete, and from the granular induration, in the albuminous deposit being of a more plastic or organizable kind; in short, the pneumonic inflammation may be said in this case to have terminated in adhesion; the albuminous effusion having partially undergone organization and contraction, and thereby glueing together and hardening the aërial cellular tissue; the whole of the appearances being, consequently, the result of permanent albuminous deposit, obstructed cells, and black pulmonary matter. I believe the interlobular cellular membrane undergoes a similar change in such cases. That this morbid change may have accompanied a pneumonia in which the reparatory process was slow, or that it may have been produced by repeated attacks or accessions of acute pneumonia, I am not inclined to dispute; but I must confess myself unable to understand the proposition, when it is asserted that this grey or black induration of the lung results from chronic pneumonia. In the wet preparation on the table, taken from a woman who lately died in this hospital, of renal dropsy, and in whom I could detect no trace of ordinary tubercle, it will be seen that whilst a portion of the pulmonary tissue has nearly taken on the form of the uniform albuminous induration, this character gradually changes as we proceed from the most inflamed part; the lung merely feels harder, is imperfectly inflatable, is of a darker colour, and is gradually lost in healthy cells.

As the morbid changes described have been somewhat indiscriminately considered as forms of tubercular infiltration, it may be fairly asked—What proof have we of their inflammatory origin? It is impossible in a necessarily limited communication like the present, to state all the circumstances, which, after several years of careful inquiry, have left no doubt in my own mind. Nevertheless, in reply to such a question, I may observe, in the first place, that we occasionally have direct proofs of a previous attack of acute inflammation within the chest. 2ndly. They are pretty uniformly accompanied by the clearest evidences of former inflammation : viz., old deposit on the pleura immediately above the affected portion of lung; adhesions between the pleuræ in that situation; considerable puckering of the pleura, with accompanying contraction and diminution of size of the lung itself, sufficient in some instances to cause actual deformity of the thoracic cavity. 3rdly. They are, perhaps, more frequently than otherwise found in the middle and inferior lobes of the lung, and not in the apices, the usual and earliest seat of tubercle; and, lastly, The total absence in many instances of a vestige of tubercle in other parts of the lung.

I cannot quit this part of my subject without adverting to the very interesting and much-agitated question regarding the relation which exists between pneumonic deposits and tubercle :—the question whether tubercle be uniformly and necessarily the result of a process of inflammation; or, if not, whether it is so in any, and in what instances. Without pretending to remove, in any considerable degree, the difficulties and obscurity which at present beset the subject, I think it desirable to place before the society a plain statement of the facts which I have observed, the analogies which have suggested themselves, and the conclusions to which I have been led by my investigations and experiments on the matter.

Like many others, I have not failed to be exceedingly struck with the general resemblance—a resemblance almost amounting to identity —observed to exist between the effects of tubercular disease occurring in serous membranes, and those met with in the lungs; the several forms, changes, and varieties presented by the one being almost equally observable in the other. The earliest and simplest form of tubercular disease in a serous membrane consists of a minute, roundish, or oval, grey, semi-transparent projection, usually hard to the touch, and possessed of considerable tenacity, so as to be broken down or disintegrated with some difficulty; being in a few instances, however, less dense, and more easily broken down by pressure. These small bodies appear to have their seat chiefly in the deeper and looser tissue of the serous membrane; sometimes, however, occupying its denser structure, being, apparently, inseparable from it and identical with it, so that an incision presents a surface in which we can distinguish no line of demarcation whatever; and, still more rarely, it would appear to be but loosely attached to the surface of the membrane, and can be separated without much violence. In like manner the earliest and simplest form presented by tubercles in the lungs is that of minute grey semi-transparent and generally hard bodies, inseparably attached to, and apparently incorporated with, the parietes of the air-cells of the lungs, in the same manner as the small tubercles are incorporated with the peritoneum. I have carefully examined these tubercles, by means of a considerable magnifying power; and am satisfied that they are not by any means distinct separate deposits, but have in reality precisely the same relation to the membrane of the air-cells as tubercles have to the cellular tissue which results from inflammation when developed in it, and as tubercles have to the membranes of the peritoneum and pleura. The best mode of showing these appearances is to inject the bronchial tubes and cells of a tuberculated lung with tallow coloured with vermilion; when not only the interlobular cellular membrane will be rendered apparent by remaining pale, but the portion of aërial cellular tissue occupied by the minute tubercles will be distinctly seen, and may be examined by means of a microscope with the aid of a reflector; the rest of the injected lobe presenting a very good example of what is called red hepatization. Should a cluster of such tubercles exist in a single lobule, and in close contact, then we have, as might be expected, a grey semi-transparent and somewhat granular state presenting itself. Neither in the case of the peritoneum, nor in that of the lung, do we at this stage discover the slightest trace of inflammatory action;—no opacity or false membrane in the former; no puckering nor condensation of adjacent cells in the latter.

We nevertheless know, from abundant facts, that the peritoneum when once tuberculated, is extremely liable to inflame; the ordinary effects of which are, opacity, thickening, and contraction of the membrane, as shown by its dull grey or whitish colour, by the diminished caliber and dense fleshy feel of the intestinal tube, by the shortening of the mesentery, and by the disfigurement of the parenchymatous viscera. These effects manifestly arise from an effusion of coagulable lymph or organizable albumen; this giving a thin coating to the serous membrane, and, by its contraction during organization, occasioning the contraction. I find in the lungs a state scarcely distinguishable from this. In certain lobules, or in a single lobule, we find a cluster of tubercles, the whole presenting now a more dull or opaque appearance: not only do they become dull and opaque, but this change is followed by a decided attempt at contraction or puckering of the affected lobule, as I believe, from a cause similar to that in the peritoneum. If we examine several of these clusters of grey tubercles in different parts of a lung, we are led to the conclusion that they begin in a point or cell, and gradually extend to the rest of the cells, either of that or of that and adjacent lobules; and that contraction first commences in the centre of the cluster or collection: so that whilst the centre has become remarkably dense, and the tubercles, as it were, squeezed up together, the density gradually diminishes, and the tubercular changes become more and more distinct, until at the very circumference we find what are called tubercles occupying a portion of pulmonary tissue still more or less healthy and crepitating. It is not a little singular, too, that as a general rule, in proportion to the degree of contraction and density of these simple tubercles, or rather, perhaps, of the film of inflammatory effusion poured out around them, is the quantity of black matter found in the diseased tissue; so that on cutting through a portion of lung so affected, the gradually diminishing density of the deposit, on the one hand, and the gradually fading black discoloration, on the other, give to the cut surface a peculiar radiating or stellated appearance. It is also worthy of remark, that in the tubercular state of the peritoneum, a collection of such black matter is by no means uncommonly met with, in small quantity, adherent both to the tubercles themselves and to the false membrane which so usually accompanies them. Admitting my description of the morbid changes to be correct, an objection may be taken to the order of succession which I have adopted. It may be said that inflammatory albuminous deposit takes place first, and that tubercles are subsequently formed in it, as manifestly often happens in false membranes, both of the peritoneum and pleura. I have very little doubt that this does happen occasionally in the lungs, as well as on the serous membranes alluded to, although it is perhaps more difficult of proof; nevertheless the reasons already given satisfy me that in general the tubercular development is primary, the inflammatory effusion and contraction secondary. On examining advanced cases of tuberculated peritoneum, I have found tuberculated parts becoming opaque, yellowish, soft, and almost like cheese: in the lungs I have found precisely similar appearances. The softening and disorganizing process, however, so commonly met with in the lungs, is less manifest in the peritoneum: the difference of the position of the two structures, and the very different degree of excitement and irritation to which they are respectively exposed, will go far to account for this; whilst it must be acknowledged that we are very far indeed from being without examples of even this change. In aggravated cases of strumous peritonitis, and especially when the intestines have been all matted together, with considerable inflammatory deposits between their folds, we not unfrequently find false membranes with every shade of tubercular change, from the semitransparent development, through all the various degrees of opacity and softening, to a yellow puriform matter, with ulceration into the intestinal tube, and consequent fæcal abscess. Again, when acute inflammation has supervened upon a tuberculated peritoneum, I have found, after death, a moderately healthy-looking plastic lymph thrown out upon the false membranes and other parts, a great portion of which manifested a strong tendency to take on a sort of tubercular form, or rather to arrange itself in separate rounded or oval masses, very closely resembling the opaque yellow soft masses occasionally met with in tuberculated lungs after a casual attack of inflammation.

As the subject of tubercle is not intended to form any necessary part of the present communication, I have merely alluded to it incidentally, in order to trace the close analogy existing between the pathological changes in serous membranes, and those of the aërial cellular tissue of the lungs; but I may nevertheless be permitted to observe, in reference to this most interesting and important question, that I fail to discover what I had always been taught to consider as essentially tubercle—a distinct, separate, or rather separable body, of a particular colour and consistence, imbedded in, and, although adhering to, supplanting a portion of the ordinary tissue of the lung. On the contrary, unless the simple transparent tubercle already alluded to can be considered as a separate and distinct body, there is not one of the varied morbid conditions, coming under the denomination of tubercle, which has not appeared to me to result from changes in or on the natural tissues, rather than from any separate and well-defined deposit displacing these tissues. These morbid changes have appeared to me to be perfectly identical with those of inflammation.

Without taking into account a highly injected state of the vessels of the part, or that arrest of secretion said to attend the onset of inflammation in every structure, the ordinary morbid changes immediately produced by pneumonia are, 1. An effusion into the air-cells of serum more or less highly charged with albumen in solution; 2. A molecular change in the inflamed tissue, characterised by more or less opacity and loss of cohesion; 3. A deposition of albumen into the cells, either solid and organizable, or fluid and puriform; 4. Total albuminization of the tissues, either in the form of a material susceptible of organization, or of a material unsusceptible of organization, and thence forming an abscess. On the other hand, the salutary or reparatory changes are, 1. Absorption of the effused serum; 2. Such a change in the molecular condition of the tissue as restores to it its natural cohesion, and, to a certain extent, its transparency; 3. Organization of the effused albuminous matter, with consequent contraction and induration of the tissues into which it is effused, or absorption of the albuminous effusion if puriform; 4. Organization of the albuminized tissue, when susceptible of that change; or the formation of a cyst to circumscribe the abscess, when the conversion is of the puriform kind. The immediate morbid changes produced by ordinary pneumonia and by phthisical disease are the same, with the exception of the albumen, whether effused or resulting from the albuminization of the tissues, being much more susceptible of organization, and consequently more likely to become permanent in the former than in the latter; whereas the reparatory or salutary changes are much less complete and much less permanent in phthisical disease than in ordinary pneumonia. Accordingly, as far as my observation has extended, the uniform albuminous induration can scarcely ever be said to have occurred in strongly-marked phthisical disease; whilst the granular induration may be said to form a connecting link between phthisical disease and ordinary pneumonia. The grey induration in phthisical subjects being chiefly made up of the less organizable or granular deposit and black pulmonary matter, is much less firm, much less permanent, and generally soon softens down into a vomica or abscess; in short, as we speak of scrofulous peritonitis, so, if called upon to give an expressive name to tubercular phthisis, I should venture to designate the disease scrofulous pneumonia. In both diseases, all the inflammatory changes may take place without either local or general symptoms sufficient to attract particular attention; in both, the general and local symptoms may be moderate, with occasional aggravations, and in both the general and local symptoms may be those of the most acute inflammation. The scrofulous pneumonia, like scrofulous peritonitis, may or may not be preceded or accompanied by the passive or simple transparent tubercle.

Some stress, indeed, has been laid upon the scattered and detached form of tubercular disease as contrasted with the more continuous and extended character of pneumonic change. I, however, believe that pneumonia itself, like tubercular change, always commences at a point in separate lobules, the continuity and extent of the changes usually found after death from that disease resulting from the rapid manner in which ordinary acute inflammation spreads; for in many cases of recent acute pneumonia we can distinctly recognise the separate points or centres: whilst, in not a few, these several centres never coalesce at all, but remain permanently distinct, and constitute what has been called lobular pneumonia. In like manner, scrofulous pneumonia, or inflammatory tubercle, whether it commence around small transparent granulations, or whether it take place in the absence of these bodies, also commences in separate lobules, which either remain permanently separated, or a large number of neighbouring lobules becoming affected at the same time, or in succession, produce a more or less continuous and extensive change, precisely analogous to, if not perfectly identical with, the effects of acknowledged pneumonia. In both the detached and continuous forms of tubercular disease just named, the hardness, the contraction or puckering, and the concentration of black pulmonary matter, sufficiently attest the attempt at reparation; although, as we know to our cost, the reparatory effort rarely proves either complete or permanent, but sooner or later gives way to softening, disorganization, and phthisical destruction of life.

It has also been made a plausible objection to the inflammatory origin of tubercular disease, that it most frequently takes place in the apices and upper portions of the lungs, whereas the reverse holds true in regard to ordinary pneumonia. Without dwelling upon the many exceptions to this rule, and without attempting to account for the greater frequency of ordinary pneumonia in the lower portions of the organs, on the score of these portions being the most dependent, and consequently most favorable to congestion, I may, without disputing their accuracy, remark, as a set-off against these statements, that it is quite in keeping with my own experience, in examining the lungs of individuals who have died of disease not affecting the chest at all, to find indications of partial inflammation much more frequently towards the apices than in the lower portions of the lungs. Partial pleuritic adhesion and pneumonic changes, according to my experience, are more frequently met with in the former than in the latter situation. If this observation be correct, will it not go some way to account for the earlier development of tubercular disease in the apices, as a general rule?

We know that the several permanent morbid structures or deposits alluded to in this paper, whether of a tubercular or pneumonic origin, maintain what may be called their integrity by a very slender power: that degree of vital influence which holds their molecules in a solid state is so inconsiderable, that it is liable to be impaired or destroyed by very slight disturbing or devitalizing causes. The most common and most powerful of these are, unquestionably, inflammation set up in the surrounding parts, and a cachectic habit of body,[1] but more especially these two causes occurring in the same individual. We accordingly find that under such circumstances these structures and deposits soften down, and are gradually converted into an albuminous or puriform fluid, and thus give rise to symptoms of phthisis. It is only with the softening of pneumonic deposits that we have to do; although, for reasons stated, it is almost impossible entirely to lose sight of the changes, which take place in tubercular deposits, and especially so as the latter are often associated with what I have designated the granular pneumonic induration.

In cases of *uniform* induration, the appearances in the lung produced by its softening necessarily vary with its seat, its extent, its being accumulated in a single well-defined mass, or distributed more or less extensively in more situations than one, and according to the completeness or incompleteness of the softening process, or the entire or partial discharge of the softened matter by the bronchial tubes. We occasionally find one large round or oval cavity, generally but not

[1] See the admirable description given of one form of such cachexy in the valuable work of Sir James Clark.

necessarily lined, or rather bounded by an opaque and sometimes moderately thick membrane, probably containing little more than a very small quantity of muco-puriform fluid, not a vestige of either pneumonic or tubercular deposit being discovered in any other part of the lungs, which shall appear to be perfectly healthy. In other instances we find the multilocular cavity or several adjacent cavities, presenting nearly the same characters, but frequently communicating with one another by bronchial tubes, the latter being generally more or less hypertrophied, and partially dilated. These multilocular cavities have, I believe, been regarded as merely dilated bronchial tubes; but I think that there can be little doubt of their having the origin I have here assigned, inasmuch as we occasionally have an opportunity of examining a lung whilst the changes described are actually in progress, as was found in the case from which the drawings were taken. In one there existed several excavations somewhat like a magnified honeycomb, completely evacuated, and lined by an opaque membrane, all appearance of pulmonary tissue being gone, whilst others were imperfectly discharged, one apparently in the first stage of softening, the grey matter being partly converted into a more yellow and more puriform-looking mass. The obscure, lingering, and less distressing form of phthisis resulting from these changes may, I think, in some instances be at least strongly suspected, if not positively declared before death, whilst not a few, I believe, completely recover to die of some other disorder.

This form of pneumonic phthisis is nevertheless much more rare than that which results from the granular deposit. The latter generally involves a great number of separate lobules, the interlobular cellular membrane remaining more or less distinguishable; and are either heaped together in a large and extensive mass, or scattered through an entire lobe, or even an entire lung. In these cases the softening process may be proceeding at a great number of points at once, and by no means necessarily commencing in the centre of each lobule; the irregular cavities thereby produced presenting, in many instances, little or none of the smooth lining membrane found in the cavities previously described, and the parietes, consisting simply of an irregularly ulcerated pulmonary tissue. It is this form of pneumonic induration which is so closely allied to tubercular or scrofulous disease of the lungs; and which, for the reasons already given, is not in itself distinguishable from it. Phthisis from this species of softening is, I believe, generally fatal; unless, indeed, it be very circumscribed, as, for example, in one of the apices of the lung.

Respecting the grey pneumonic induration, I have little to add to what has been already stated. The softening usually takes place very slowly, either in one or in several points: the cavities thence resulting are, I believe, more frequently of moderate than of large size; and unless secondary changes and deposits shall have taken place from accidental attacks of inflammation, the resulting phthisis is proportionably slow. In the few well-marked cases I have seen the disease proved fatal through the supervention of a more or less acute attack of bronchitis or pneumonia.

In regard to the question respecting the cicatrizing of a vomica in the lung, I may observe that the uniform albuminous, as well as the grey induration, is frequently distributed in a sort of linear form, so as very much to resemble a cicatrix; and considering the perfect identity of the physical signs of pneumonic consolidation with mucus in the tubes of the consolidated part, and those of pneumonic abscess, on the one hand, and the linear distribution of pneumonic deposit, with consequent puckering of the lung on the other, I cannot resist the belief that many at least, if not all the reputed cases of cicatrized abscess, may have been of this kind; although I advance the suggestion with all the deference due to the high authority of Laennec, and the scarcely less valuable opinion of Dr. Stokes.

Having described the ordinary deposits, diffuse and lobular, produced by acute pneumonia; and having stated what I believe to be the ordinary changes which take place in them in an after period, I may be permitted, in conclusion, to refer to the changes which we so often have an opportunity of seeing in the apex of a single lung, as furnishing a very good illustration and sort of epitome of the whole of this part of our subject. The appearances alluded to are frequently discovered after death, without the least suspicion of their existence having been previously entertained, being merely incidental, and often met with in the bodies of those who have died a violent death, and who might be said to have been previously in perfect health. The total absence of tubercles, or any other disease whatever, in any part of either lung, goes far to negative their tubercular origin: the thickened pleura pulmonalis, the adhesion between the pleura pulmonalis and pleura costalis, together with the irregular puckering of the lung and pleura pulmonalis, sufficiently attest the previous existence of inflammation. On cutting into the part we sometimes find a specimen of the uniform albuminous induration from acute pneumonia; or, though much more rarely, a well-defined round or oval cavity, lined by a smooth membrane. In other cases we have a very good example of the granular and yellowish deposit, probably in different stages of change, firm and dry, or softening down in various points, or forming irregular or tortuous sinuses or cavities having an uneven, naked, or ulcerated surface; or, here and there, portions of dark-grey indurated lung puckered up, and perhaps embracing an irregular earthy concretion, thus constituting a mixture of the granular and iron-grey induration. In another case we find it to consist entirely of the iron-grey induration, with or without partial softening or earthy change. Lastly, we may find the uniform albuminous, or the grey induration, so arranged, and occasioning such puckering of the pleura and lung, as strongly to suggest the notion of a cicatrix.

PLATE I.

a. Portion of the superior lobe of the lung in a healthy state.

b. An irregular cavity, from softening of light-grey induration.

c. A large cavity, evidently of long standing, lined by a smooth firm membrane, and communicating with a large bronchial tube, into which a bristle is inserted.

dd. Sloughy softening of grey induration, bounded by a membrane, as seen at section *e.*

f. Uniform albuminous induration in its entire state, its contraction having led to a marked diminution in the size of the lung.

g. Septum between the lobes of the lung.

h. Old albuminous deposit on the pleura, situated above the large excavation.

PLATE II.

Represents a section of lung affected with partial iron-grey induration.

a a. Pleura pulmonalis, thickened, contracted, and corrugated by old albuminous deposit.

b. Dark iron-grey induration of the lung, gradually diminishing from a centre, so as at length to disappear in healthy pulmonary tissue; which is, however, rather more than usually charged with black matter.

c. Portion of iron-grey induration, softened down into a vomica, and circumscribed by an albuminous deposit, as indicated by the surrounding white line.

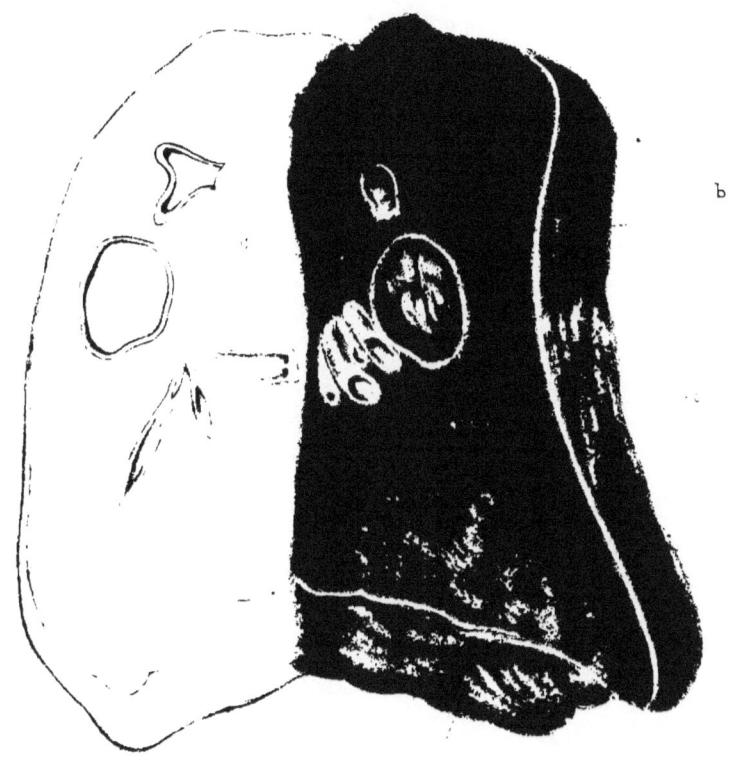

a.

b

W Hurst del. W West Lith.

PLATE III.

Fig. 1.

Represents a portion of lung affected with uniform albuminous, and light grey induration, softening down into cavities.

a. Uniform albuminous induration in progress of softening.

b. Light grey induration undergoing a similar change.

c. Cavities remaining after the expectoration of the softened tissues.

Fig. 2.

Portion of lung affected with iron-grey induration.

a. Cavity resulting from softened grey induration.

b. Bronchial tube and branches extending to masses of calcareous matter, the residue of softened grey induration.

Fig. 3.

Represents softening of grey induration, as shown at $m\,m\,m$, communicating with the bronchial tubes, as shown at n. o. A bloodvessel traversing the softened tissue, its parietes thickened, and its distal extremity closed in a conical form by the contraction of surrounding albuminous deposit.

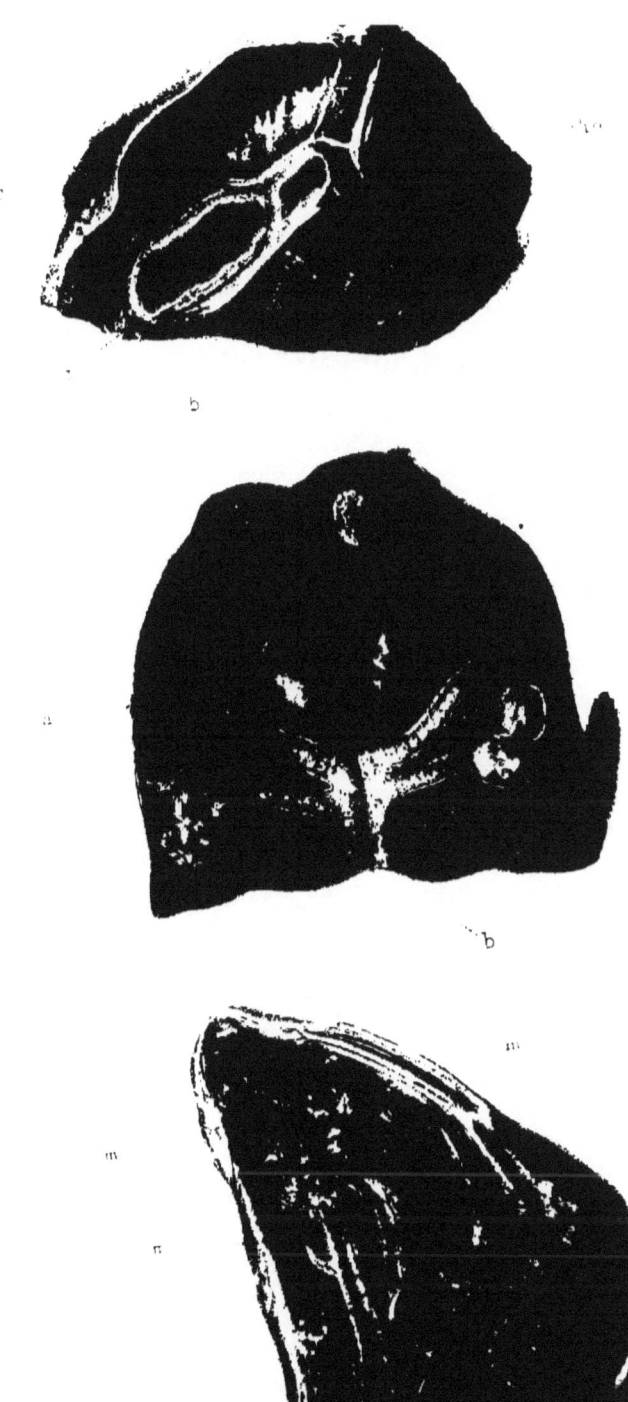

ON

THE PATHOLOGY OF PHTHISIS.

Read before the Guy's Physical Society, Jan. 4th, 1845.

AT pages 237 and 238 of the 'Elements of the Practice of Medicine,' published in 1837, the following passage occurs in relation to changes produced in the lungs by pneumonia:—"In some instances, however, when the albuminous matter thrown out is of the more plastic or organizable kind, it fails to be entirely absorbed, and part of it permanently remains. Under these circumstances we find it, at an after period, either in small, detached, and more or less rounded masses; or more extensively and more irregularly diffused through the pulmonary tissue. When distributed in small insulated portions it constitutes one of the forms of albuminous deposit, indiscriminately called tubercles; whereas, when more extensively and irregularly diffused, it has, in like manner, been regarded as a form of tubercular infiltration. The history, however, of the patient's case, in many instances, as well as the local appearances themselves, lead to the conclusion that they are merely the result of a previous attack of pneumonia. We often learn, on inquiry, that at some former period, perhaps years before, the patient had had an attack of inflammation within the chest; whilst if he die of some other disease we almost uniformly discover, on dissection, unequivocal evidence of antecedent inflammation. The evidence consists in thickening and adhesions of the pleuræ, especially in the neighbourhood of the appearances in question, together with induration and puckering of the pulmonary tissue immediately surrounding each albuminous deposit; or, when the deposit is irregular and extensive, we often find an actual deformity and puckering of the pleuræ above the infiltrated parts. This view of the origin of these albuminous deposits will probably serve, in some measure, to explain why they are much less uniformly found in the apices of the lungs than ordinary tubercles. It has been observed that these deposits may remain passive for an unlimited period, and without undergoing any very appreciable change, except, perhaps, a conversion of some of them into calcareous or chalky masses, especially when deposited in the upper lobe of the lung: it would nevertheless appear that the vital influence by which they are maintained in their integrity is so extremely slender, that if inflammation happen to be set up around them by any accidental cause, and especially if the vital powers of the patient have been greatly impaired, that influence is so far exhausted, that they lose their cohesion and soften, &c."

In a paper "On the Anatomy of the Lungs," read before the Royal Medical and Chirurgical Society, April 23, 1841, amongst other propositions will be found the following:—" Pneumonia and inflammatory tubercle are identical." "Ordinary tubercles present the same varieties in the lungs as they do in serous membranes." Ever since the above statements were made public, and long before, I have omitted no opportunity of seeking to test them by a constant attendance at the inspection of bodies made at this hospital, and by a careful and often repeated examination of morbid parts made at my own private residence.

The subject, as expressed in the above quotations, suggested for consideration two very important but perfectly distinct questions:—
1. The pathology of pneumonia. 2. The pathology of phthisical disease; but more especially those forms of it which occur in connection with pulmonary tubercle.

In venturing to place these two questions more in detail before the profession, I could not but feel that the former, or the pathology of pneumonia, ought, as a matter of course, to take precedence of the latter—the pathology of pulmonary tubercle—inasmuch as it appeared to me an almost hopeless task to attempt any analysis of the pathological appearances met with in the lungs of persons dying of phthisis, without a previous knowledge of the changes produced in the pulmonary tissue by the mere process of inflammation. Accordingly, in my last communication, my chief object was to trace the morbid changes induced by pneumonia; and, more especially, to demonstrate the fact that that disease not unfrequently leaves behind it divers forms of induration of the pulmonary tissue, all of which have been somewhat indiscriminately, but, I believe, erroneously, regarded as modifications of tubercle or of tubercular infiltration. Such indurations of the pulmonary tissue, altogether independent of tubercle, are, I am more and more persuaded, of very common occurrence; and moreover venture to predict that the time is not far distant when it will be generally acknowledged that they have not by any means commanded the attention they deserve.

The comparative neglect, hitherto, of these pathological changes is, doubtless, in a great measure, owing to the erroneous impression left by the illustrious Laennec respecting the signs and symptoms of what he was led to regard as the simplest form of ordinary pneumonia. In a former paper I endeavoured to show that the simple pneumonia of Laennec is in reality a broncho-pneumonia, and that there exists a still more simple form of the disease incident to all ages and constitutions, in which there is but little cough or expectoration, and in some instances none at all, and that, in consequence, this simpler form is often entirely overlooked. It may be true that this simpler form of pneumonia is most frequently met with in the aged and in cachectic habits, constituting what have been called the latent and typhoid forms of the disorder; but it is nevertheless a very great and very serious error to conclude that it is confined to such persons and constitutions. Nor do I suppose that there is a single experienced physician present who has not had more opportunities than one of satisfying himself that such cases, occurring even in early life and in hale constitutions, are very liable indeed to be either altogether overlooked, or mistaken for, and treated as cases of common continued fever.

If, then, this simpler form of pneumonia be liable to escape detection, even in early and middle life, and in good constitutions, how much more likely is this to happen when it occurs in the aged and in highly scrofulous or otherwise cachectic habits. It cannot be doubted that simple acute pneumonia occurring in such persons and constitutions must be still more liable to be overlooked or mistaken, inasmuch as neither the constitutional symptoms nor physical signs are in general so unequivocally developed. This is so true that even the crepitating rhonchus, so much and so justly relied upon in the diagnosis of ordinary cases, is often exceedingly transient or imperfect, as if the rapid congestion and consolidation of the pulmonary tissue excluded that degree of serous infiltration which is, apparently, essential to produce crepitation. When we take into consideration these sources of error, against which I know that even the accomplished stethoscopist is not universally or altogether proof; and when to this we add the fact that a very considerable proportion of the profession still unfortunately continue to disregard auscultation, or are unable to avail themselves of its advantages, we cannot be surprised that morbid changes, the result of a bygone inflammation of the lungs, should frequently fail to be recognised during life. But I venture to carry the argument further, and to submit that acute pneumonia, in its most ordinary forms, and occurring in the best of constitutions, and unequivocally recognised at the time, not unfrequently leads to the permanent indurations I have described, without even the practised stethoscopist being able positively to satisfy himself of the fact. The explanation I believe to be this: in such cases pleurisy has co-existed with the pneumonia; the physical signs which might otherwise have proclaimed the presence of consolidation have been masked by the pleuritic effusion; and the permanent dulness of sound on percussion, absence of respiratory murmur, rough tubular respiration, and modified bronchophony, have been attributed to the presence of the false membranes resulting from organization of the albuminous portion of that effusion.

Another cause of these permanent indurations of the lung having been so much disregarded or overlooked, is to be found in the very vague and I believe erroneous opinions that have been entertained respecting tubercle, and what has been called tubercular infiltration. In short, I suspect the history of these inflammatory indurations will prove to be very analogous to that of pericarditis. The latter disease was at one time supposed to be extremely rare, and almost uniformly fatal; whereas we now know that both propositions are equally unfounded: so, in like manner, when acute pneumonia proves immediately fatal, these pneumonic indurations necessarily cannot exist; the hepatization which led to them is, for the reasons stated, overlooked; and hence, when discovered at an after period, they are at once set down to the score of tubercle, or tubercular infiltration, according to the circumscribed or diffused form which they may happen to have assumed;—an error long promoted and apparently confirmed, by the fact of these indurations being, beyond all dispute, most frequently met with in tuberculated lungs. But whether they occur in tuberculated lungs, or in lungs unoccupied by tubercles, they have the same origin: they are the result of pulmonic inflammation: they present the same changes in their progress to permanent induration or repair, and manifest the same tendency to retrograde or disorganize, at various periods of their existence, according to the state of the patient's constitution, or under the influence of certain morbific causes acting more immediately upon the lungs themselves: they may soften down or even form sloughs, at a very early period of their development or formation, perhaps a few weeks or months after the occurrence of the inflammation which gave rise to them; the pulmonary tissue, although consolidated, still presenting various degrees of brittleness, or loss of cohesion; or, they shall proceed to perfect hardness and quiescence, and either remain passive and without creating any inconvenience, or from some cause or other, after an indefinite period, begin to disintegrate and give rise to excavation, with most of the ordinary signs and symptoms of phthisis.

During the last eight or ten years I have had opportunities of distinctly tracing the progressive changes alluded to in lungs unoccupied by a single tubercle; and in cases where the pneumonic inflammation giving rise to them had been satisfactorily recognised during the life of the patient. Such instances, beyond all doubt, occur most frequently, but not exclusively, in persons of a bad or cachectic habit of body; whilst I very much question whether there ever was a single instance of tubercular disease of the lung proving fatal, in which more or less of this pneumonic change might not have been distinctly recognised, if the prevailing notions respecting tubercular infiltration had not obscured the perception of the beholder.

It is a curious and highly interesting question to the pathologist how far the recent pneumonic change must, in any instance, have proceeded, to have passed beyond the limits of complete restoration of the normal condition of the pulmonary tissue. Undoubtedly this will depend very much upon the condition and constitution of the patient; permanent albuminous deposits being much less frequently met with in hale and healthy subjects than in persons of a strumous or cachectic constitution. That red hepatization may subside, and ultimately leave the pulmonary tissue little if at all changed, is, I think, hardly to be disputed; but how far, or to what extent, an albuminous deposit, sufficient to produce grey hepatization, shall have proceeded, to admit of complete restoration, is a question by no means of easy solution. If the opacity produced by inflammatory action in a serous membrane, such as the arachnoid, pleura, or peritoneum, can be perfectly repaired, and transparency restored, it is easy to conceive that a similar—probably merely molecular—opacity, giving rise to early grey hepatization, may admit of a similar result: and from my own observation I am led to believe that it is so. But whether the more decided and distinct albuminous material often met with in feeble constitutions can ever be removed, and the transparency of the pulmonary tissue restored, I confess that I am not in a condition to speak with any confidence. It may be so, and especially in moderately good constitutions; but I am in possession of no absolute proof of the fact, nor do I know how any positive proof is to be obtained. When such albuminous effusions have proceeded to actual induration, and consequent obliteration of the cellular structure of the lung, the notion of perfect repair and restoration can hardly, I think, be regarded in any other light than that of a pathological absurdity, although the statements advanced by some modern writers on phthisis would seem to imply that they actually expected the profession to believe in its practicability.

If I shall succeed in directing more precise attention to this subject, I cannot entertain a doubt that some practical good will result; and more especially that the knowledge thus acquired will be brought to bear upon the important question of diagnosis in phthisical disease;—a question, the satisfactory solution of which is calculated, above all things, to extinguish the quackery and silence the reckless pretensions which from time to time are brought forward to insult the understanding and shock the better feelings of every upright and really competent member of the profession.

As it is not my intention to enter upon the subject of diagnosis in my present communication, I shall content myself with observing that the permanent indurations of the pulmonary tissue which I have described, especially when partial, are occasionally accompanied by considerable dilatation of the bronchial tube or tubes passing through or near them. These dilatations I believe to be an effect, and not a cause of the indurations, as some have imagined. This state of things—induration with dilated tubes—may occur in any part of the lungs; and I shall briefly relate one or two cases in illustration.

A girl, in Charity Ward, was exceedingly ill, but the details of her illness it is unnecessary to dwell upon; suffice it to say, that on examining the chest, I observed towards the inferior and anterior part of the right side near the sternum some dulness of sound on percussion, to which I should probably have attached little importance if I had not at the same time detected bronchophony and tubular respiration. Discovering no physical indications of phthisis elsewhere, I inquired if she had had whooping-cough, and in reply was told that she had had the disease severely some years before. She died and was examined, and, as had been anticipated, she had evidently been the subject of pleuro-pneumonia at some former period. The pleuræ were strongly adherent at the spot indicated; the pulmonary tissue beneath was indurated, and the indurated part was traversed by a remarkably dilated bronchial tube. The preparation has been mislaid.

It is, however, when consolidation occurs at the apex of the lung, even without any considerable bronchial dilatation, that there is the greatest difficulty of diagnosis, and the almost certainty of at least suspecting it to be disorganizing phthisis, especially, if with consolidation we happen to have an excess of mucous secretion in the bronchial tubes. Under such circumstances the physical signs of phthisis are complete.

A short time ago a case was attended in this hospital conjointly by Dr. Lever and myself. To my indefatigable friend I am indebted for the following history.

"Rebecca M—, a married woman, 29 years of age, presented herself amongst the out-patients of Guy's Hospital, complaining of hæmorrhage from the uterus. She had borne several children, and had miscarried once or twice. The hæmorrhage had existed between four and five weeks, and at the time of her visit was profuse. She had placed herself under the care of some chemist, and had taken various medicines, some of which had diminished the flow of blood, but from the time of its commencement it had never entirely ceased. She was pale, exsanguineous, and exceedingly debilitated, her feet and ankles were œdematous, her pulse small, rapid, irregular, and intermitting. Five years previously she had been laid up for several weeks with 'rheumatic fever.' She stated that for three years she had suffered from a cough, the commencement of such being an attack of inflammation of the lungs; that she expectorated a little brownish-green matter; and that she occasionally suffered from great difficulty of breathing; but that her cough had been less, and her breathing more easy, since the occurrence of the flooding.

"There was a sensation of weight at the bottom of the back, but there were no lancinating central pains which usually attend malignant organic disease of the uterus. An examination per vaginam was instituted. The uterus was found to be flabby, relaxed, and large, but free from organic lesion. The following medicine was prescribed:

Alumin. gr. xij;
Acid. Sulph. dil. ♏x;
Tinct. Hyoscy. ♏xx, t. d.
Ex. Inf. Rosæ. Comp.

This she took for three weeks with an occasional saline aperient.

"I lost sight of her for several weeks, or I may say months, when she again presented herself; complaining of amenorrhœa, cough, pain in the chest, palpitation of the heart, dyspnœa, &c. The amenorrhœa was evidently owing to pregnancy; and for the relief of her other symptoms I advised her to seek admission into the hospital; this she did, and on the following Wednesday was placed under your care.

"Two weeks after she left the hospital she aborted. Her husband being from home, she was alone when this accident occurred, and was for twelve hours unseen and unattended to. A considerable quantity of blood was lost, and for some hours her condition was precarious. (She had attained the fifth month of pregnancy.) For three days following abortion there was diarrhœa, accompanied with some pains in the bowels; tenderness of the abdomen; her cough was distressing; her dyspnœa urgent; the voice was husky; the expectoration tough, yellowish-green, and streaked with blood, and occasionally with it was mixed a considerable quantity of rusty-coloured froth; the pulse was rapid, weak, irregular, and intermitting. Soon after the irritable condition of the bowels was relieved, she suffered from profuse perspiration, occurring three or four times in the twenty-four hours, but most distressing in the night. In the course of a week the perspirations diminished, but in their place supervened the diarrhœa, and so matters continued until the period of her death, three weeks from the time of abortion. At the time of her disease she was sitting up in bed in the act of coughing, when a quantity of purulent matter mixed with blood flowed from her mouth, and she fell back and expired."

Although this patient had organic disease of the heart, and was suffering from uterine disturbance, it was believed at the time that she was also the subject of phthisis. She had cough and expectoration, with occasional hæmoptysis, and considerable emaciation. There was dulness on percussion at the right apex, with depression of the ribs, bronchial respiration, bronchophony, and mucous gurgling. Nevertheless, the history of the case and some other circumstances taken in connection with it led to the suspicion that the signs and symptoms might be the result of bronchial irritation in a lung damaged by an ancient pleuro-pneumonia. Through the kindness of the Doctor, we obtained permission to examine the body. The pleuræ were strongly adherent at the apex, and a portion of the lung beneath was in a state of passive iron-grey induration. There was no destructive disorganization whatever; nor was there the slightest trace of tubercle to be found in either lung.

Still more recently a female patient in the hospital died of dropsy. She had been under the care of one of my colleagues. She was supposed to be at the same time the subject of phthisis, all the physical signs of that disease being strongly marked at the apex of the lung. On dissection, the pleuræ were found adherent, the lung itself beneath the adhesion was in a state of well-marked iron-grey induration, there was no destructive disorganization whatever, nor was a tubercle to be found in any part of either lung.

Now that morbid state cannot be of rare occurrence of which three well-attested cases come under the observation of a single individual in the short space of a few weeks, nevertheless at the present time it is only vaguely acknowledged, and is still without a name.

The following is still more curious and interesting. It is certainly upwards of four years ago since Mr. G—, the brother of a medical man, first consulted me at the suggestion of my friend Mr. Key. He presented the ordinary signs of phthisical disease at one apex, but with indications of an unusual degree of consolidation. He was exceedingly hoarse, had cough and expectoration, some hectic and considerable emaciation. I saw him from time to time during the last few years of his life, and often had cause to be struck with the quiescent condition of the parts of the lung originally consolidated. The physical signs extended themselves progressively over a very considerable portion of the lung first affected; and at length presented themselves in the other, in the same order. After dragging on a sad and melancholy existence for four or five years in this way, he was at length assailed by peculiar affliction, and sunk. For the particulars of the inspection made a few months ago I am indebted to Mr. Foaker of Great Baddow, through the courtesy of Dr. Miller of Chelmsford.

"I now regret that no notes were made of the morbid appearance found in the corpse of poor G—; more especially as some circumstances were hardly reconcileable with each other, as you will remember. We remarked a lodgment of pus in the lower portion of the trachea, completely obstructing both bronchi, sufficient, I conceived, to have produced suffocation. Nevertheless, there was nothing like excavation in either lung, though cut into every lobe; and the cut surfaces exuded a gritty half-purulent fluid-like softening tubercle. Both lungs were hard and incompressible, when not adhering to the costal pleuræ; their coloured and speckled appearance resembling the surface of granite. There was no ulceration in either the trachea or on the mucous membrane of the small intestines, although some spots on their peritoneal coats led me to open them.

"I think I did not see him the last three or four weeks of his life, but could not learn that the sputa ever exhibited a purulent character. I am sensible how incomplete this statement is in a pathological view, and trust to your own observations and memory to supply what may be more particularly desired by your friend.

"Yours truly,
"L. FOAKER."

Who could have doubted the existence of more or less excavating disorganization in such a case as this?

From what has been advanced in this and former communications, I hope it may be permitted to arrange the several forms of disorganization of the lungs resulting from mere inflammation and its consequences, under the head of

PNEUMONIC PHTHISIS.

This pneumonic phthisis may be acute; the deposits and inflamed tissues softening down and disorganizing at once, without any attempt whatever being made at induration or repair; thereby constituting one form of acute or galloping consumption. It may be *acuto-chronic*, of which I would distinguish three varieties:—1. The inflammation, though more or less acute, is slower and more insidious in its course, and manifests some attempts at repair, as indicated by various stages and degrees of induration. The induration, nevertheless, is not complete; the pulmonary tissue continues to be friable; and sooner or later, that is to say, in a few weeks or months, softens down, and gives rise to excavation; most frequently by a somewhat slow ulcerative process; more rarely by an actual slough of greater or less portions of the indurated but still friable pulmonary tissue. 2. Inflammation may supervene upon or around ancient induration, leading to disorganization either of the newly-inflamed tissue, of the old induration itself, or of both at the same time. Lastly, pneumonic phthisis may be *chronic*, of which I would also distinguish two varieties; first, that in which old indurations undergo a slow process of disintegration, giving rise to vomicæ; and secondly, that very rare form of the disease, in which an insidious inflammation proceeds very slowly to convert a considerable portion of pulmonary tissue into grey induration, without any necessary excavation whatever, as in the case of Mr. G., before given.

TUBERCULO-PNEUMONIC PHTHISIS.

By tuberculo-pneumonic phthisis is meant a very common form of the complaint, in which, although tubercles are present, the really efficient cause of the phthisical mischief is pulmonic inflammation. In this form of the disorder the tubercles sufficiently indicate the strumous or cachectic habit of the individual, and manifestly predispose to the inflammatory change: nevertheless they do not, beyond this, seem to be either primarily or essentially concerned in the serious changes observed to take place in the pulmonary tissue.

As the time of the Society is valuable, and I have no inclination whatever to write the voluminous details which my subject would admit of, and almost justify, I have, at the risk of being considered dogmatical, resolved to state my opinions in a very brief and somewhat unconnected or aphoristical manner. It will at least be favorable to discussion.

However analogous and closely allied the abnormal nutrition which produces tubercle may be to that which constitutes inflammation, we cannot, in the present state of our knowledge, admit their identity.

The great transparency in most instances of simple pulmonary tubercle, for some time after its development, and the total absence of any degree of surrounding opacity, are irreconcilable with the necessity of actual inflammation for its production: nevertheless, pulmonary tubercles, and the ordinary effects of inflammation, have been confounded with one another.

Simple pulmonary tubercle is most frequently of a grey colour, vitreous-looking, or semi-transparent, moderately hard, offering considerable resistance to pressure between the finger and thumb, and is apparently homogeneous. It varies in size from that of a mere point, to that of a mustard-seed, and rarely exceeds the latter.

To these tubercles, by way of distinction, I would apply the term *sthenic*. They are comparatively little prone to disintegration; a tendency to which is indicated by yellowish opacity and loss of cohesion.

In other instances simple pulmonary tubercle is, from the first, and however minute, of an opaque white, or boiled-rice colour; sometimes with a faint tinge of yellow; softer and more friable than the above; and occasionally attains a somewhat larger size. To these I would apply the term *asthenic*. They are much more prone to disintegration.

Although no precise line of demarcation can be drawn between these two varieties of simple pulmonary tubercle, the one or the other is observed to preponderate in different cases of phthisis. The sthenic may, by deterioration, assume the character of the asthenic, but not the converse; the uniform tendency of all tubercles being retrograde, or towards loss of cohesion and disintegration.

It is the sthenic variety of tubercle that usually preponderates in the tuberculo-pneumonic phthisis now under consideration.

Pulmonary tubercle has its seat in the delicate filamentous tissue which forms the slight filmy parietes of the air-cells, and bears the same relation to these parietes that tubercle does to serous membrane.

The apparent growth and increase of size of simple pulmonary tubercle, beyond that specified, depend upon changes taking place in adjacent cells, either from the development of additional tubercles, or from inflammation.

The so-called enlarged tubercles, therefore, are, in reality, either aggregations of simple tubercles, or simple tubercles enclosed in the products of inflammation. To the former I would apply the term *compound tubercles*, which, of course, are made up of two distinct elements, viz., the abnormal product, tubercle, and the tissue of the air-cells in which that tubercle is developed. The latter element of compound tubercle, although little considered, probably plays a by no means unimportant part in some of the most serious changes observed to take place in tuberculated lungs.

Compound are at all times much more liable to disintegration than simple tubercles.

The abnormal hypernutrition which leads to tubercle appears to be connected with a scrofulous habit.

When tubercles are present, and especially when numerous, in clusters or compounded, but still quiescent, the neighbouring pulmonary cells pretty uniformly afford indication of compensatory or excessive function, the cells being more or less enlarged, as observed in pulmonary emphysema.

This compensatory change probably increases the difficulty of recognising simple tubercles in the lungs, it being rarely, if ever, possible to detect them, either by dulness of sound on percussion, or by diminished respiratory murmur as determined by auscultation, unless, perhaps, they exist in great numbers and in groups of considerable size. A case which recently occurred at this hospital very well illustrated this position. A little boy was emaciating and gradually sinking under a severe attack of what we have now no doubt was infantile remittent fever. From the obscurity of the case, our attention was particularly directed to the lungs, in order to ascertain if any phthisical disease existed; but although he was repeatedly and carefully examined by several gentlemen, with the express view of detecting tubercles, nothing in the least satisfactory resulted. Nevertheless, after death, simple tubercles, in great numbers, existed at the summit of both lungs, as may be seen by inspecting the preparation placed upon the table.

In general, the only physical indication worthy of the smallest confidence at a very early period is a certain degree of inequality in the respiratory murmur, it being a little more loud or puerile at certain points, or in one lung than in the other.

Having no means of positively ascertaining the presence of simple tubercles, we are unable to determine either their first development, how rapidly they are developed, or how long they may remain in a passive or harmless state.

When tubercles are present in the pulmonary tissue there exists a great proneness to increased vascular action, congestion, or inflammation in the lungs themselves and their appendages, the larynx, trachea, bronchi, bronchial glands, and pleuræ. The tendency is, however, most strongly marked in the immediate vicinity of the tubercles.

Unless subjected to causes calculated to aggravate such tendency to congestion or inflammation, an individual may experience little or no inconvenience from the presence of tubercles—a pathological principle applicable to sthenic tubercles generally, whether situated in the lungs, the pleuræ, peritoneum, or arachnoid.

The inflammation which supervenes upon tuberculated lungs is commonly of the low insidious kind observed in scrofulous and cachectic constitutions; and, unless under aggravation, is rarely attended with sufficient serous exhalation to produce perfect pneumonic crepitation.

This inflammatory condition of the lung is very commonly accompanied by a corresponding condition of greater or less portions of the mucous membrane of the air-passages; and especially of the neighbouring bronchi, which, under the circumstances, are usually found somewhat dilated, and have their parietes thicker and softer than natural.

The commencement of this inflammation, either in the pulmonary tissue itself or in the bronchial tubes, is the ordinary commencement of tuberculo-pneumonic phthisis; and hence it is that most of the symptoms and physical signs of phthisis may manifest themselves without evidence of a primary change in the tubercles themselves being discoverable after death.

When the pulmonic inflammation is more considerable than usual the characteristic heat of skin often remains uninterruptedly for days together, whether accompanied by diminished, increased, or varying perspiration.

The physical signs which become apparent as the disease advances, viz. feebleness or absence of respiratory murmur, bronchophony, tubular respiration, and dulness of sound on percussion, are more the results of this inflammation than of the extension of tubercles; and may occur from a similar cause, in pneumonic phthisis, without a tubercle ever having existed in the lung at all.

There is often a well-marked relation to be observed between the degree and permanency of the characteristic pneumonic heat of skin when present, and the rapidity and extent of these local changes, as ascertained by auscultation and percussion.

The inflammatory changes which take place in tuberculated lungs resemble those resulting from inflammation of the lungs without tubercles.

Although the red hepatization occurring in tuberculo-pneumonic phthisis very often passes quickly into softening, and the consequent formation of a cavity, nevertheless, when actual albuminous matter is thrown out, it, like that resulting from pulmonic inflammation without tubercle, usually manifests some attempts at repair, as indicated by more or less hardening and contraction of the deposit itself, and of the pulmonary tissue, into which it is effused.

These results of inflammation have been very commonly, but erroneously, regarded as mere varieties of tubercular infiltration.

It is the contraction of these deposits that chiefly occasions the diminished size of the lung and flattening of the ribs. Pleurisy, doubtless, very often contributes to the same result; but mere excavation of a portion of a lung has no such effect as diminishing its size. In the worst and most excavating forms of emphysema there is no shrinking of the lung.

The same contraction of albuminous matter on exposed blood-vessels acts as a ligature or roller, helps to obliterate their canals, and so to prevent hæmorrhage. Nevertheless, the parietes of an exposed blood-vessel do occasionally soften or ulcerate, and thus give rise to the serious hæmorrhages of advanced phthisis.

The attempts at repair in the pneumonic deposits which occur in tuberculo-pneumonic phthisis lead most frequently to the iron-grey and granular indurations pointed out in a former paper. In those rare instances where the character of the pneumonic induration approaches that described under the name of uniform albuminous induration, its semi-transparency occasionally imparts to it a gelatinous or horny aspect.

When the deposit and subsequent induration are limited to the vicinity of a simple tubercle, or are otherwise very cicrumscribed, we occasionally find, as the result, a small, irregular, hard, and usually dark-coloured body, which is firmly adherent to the more healthy tissue.

The attempts at repair, or induration of the pneumonic deposits occurring in tuberculo-pneumonic phthisis are commonly very imperfect, and not durable; so that the deposits, for the most part, sooner or later undergo a second change, by which they soften down, and produce excavation. This softening, however, may take place days, weeks, months, or, I believe, even years after the original deposition.

When the disease has proceeded to excavation, the natural cure of the ulcer thus produced consists in the formation of a more or less dense and paramount lining membrane—the true cicatrix of such ulcers.

Although this membrane may perhaps now and then remain passive and harmless for years, it most commonly happens that the cicatrization is imperfect and incomplete; the efforts at repair fail; the albuminous material, which ought to form the membrane, softens down, and with it successive portions of the pulmonary tissue furnishing it; and thus the ulceration proceeds till, exhausted by unceasing irritation and imperfect nutrition, the patient dies.

In examining the lungs of fatal tuberculo-pneumonic phthisis, we not unfrequently have an opportunity of observing, along with or without disintegrating tubercles, all the different forms or stages of the pneumonic process:—a dull red and congested state of the cells; diminution of cohesion; red hepatization; red hepatization at once softening down, forming cavities containing a dirty puriform fluid; grey hepatization; grey hepatization softening down into cavities; albuminization of portions of the lung; various stages or degrees of hardening and contraction of the uniform, grey, or granular induration; secondary softening of one or more of the latter; together with more or less dilatation, thickening, softening, and opacity of the bronchial tubes passing through the diseased parts; such bronchial tubes very commonly presenting, on being cut across, an appearance which has often been mistaken for tubercle, softening in the centre. It is only, however, in the very worst forms of tuberculo-pneumonic phthisis that we meet with all these appearances;—cases approaching more nearly to what I shall have to describe, under the name of *tubercular phthisis*.

In some instances there seems to be no attempt whatever made at repair in the pneumonic deposits; instead of which, they, as well as the tissue into which they are effused, soften down at once, producing one form of what has been called galloping consumption. This result, however, is chiefly met with in the more asthenic disease —tubercular phthisis.

When, on the other hand, diffuse inflammation attacks lungs studded generally with simple tubercles, it gives rise to what I would call *suffocative pneumonia*; characterised more particularly by extreme, or at least disproportionate dyspnœa, lividity of lips and countenance, and very commonly by the rapid dissolution of the patient. In this case the inflammation usually proves fatal before albuminous deposits occur at all, and more or less crepitation is heard till the last.

Although the softening of ancient pneumonic indurations appears to be most frequently brought about by inflammation accidentally set up around them; a cachectic or bad habit of body, however induced, tends to a similar result; not, probably, by actually predisposing to inflammation, but rather by lessening the tone of the inflamed or irritated tissue, and so accelerating disorganization or disintegration.

It is the softening of these indurations, or of the hepatizations which precede them, that constitutes the principal source of disorganization and excavation in tuberculo-pneumonic phthisis.

When the pneumonic deposits have proceeded to that state which constitutes either the uniform albuminous, the iron-grey, or the granular induration, although they may remain in a passive and harmless state for an indefinite period; it is impossible to imagine that they can ever be removed, and the natural and healthy tissue of the lung restored.

The amount of expectoration depends upon the extent to which the bronchial tubes happen to be implicated.

The appearance of the matter expectorated affords, in a large majority of instances, no positive test, either of the existence or of the extent of disorganization.

Possessing no means of removing tubercles when once developed, our resources are purely preventive; and are such as are calculated to improve, strengthen, and fortify the general constitution.

Our next duty is, as far as is consistent with the above principle, to remove all the predisposing, and avoid all the exciting causes of inflammation of the lungs and their appendages.

Tuberculo-pneumonic phthisis having once fairly commenced, the inflammatory process must be met and combated by moderate general or local bleeding, active and continued counter-irritation, the general use of antiphlogistic measures generally, bland diet, warm clothing, regulated temperature, or, what is infinitely better, a milder and less variable climate.

When considerable, as indicated by the enduring pungent heat of skin, increased hurry of respiration, sharpness and hardness of the pulse, and by a small mucous or subcrepitant rhonchus, the accessions or aggravations of pulmonic inflammation may, to a certain extent, be checked or controlled by the use of mercury in addition to other means; but when mercury is made to affect the general constitution, it has appeared, in some instances, to hasten disorganization, at a subsequent period.

The disease, by proper treatment and management, may often be arrested for a time, and not very unfrequently for years, the pneumonic changes already produced remaining in a passive state, as determined by auscultation and percussion.

It is but just, and of some importance to admit, that although the general tendency to repair or induration in pneumonic deposits, and the absence of it in tubercle, are strongly marked, and sufficiently distinguishable, it is sometimes difficult or impossible to pronounce with certainty whether the change in the pulmonary tissue be the result of tubercles or of a bygone inflammation; neither is it at all times easy to decide whether the inflammatory changes, when sufficiently evident, have or have not been preceded by tubercle. It is not improbable, however, that injection of the blood-vessels, organic chemistry, and the use of the microscope, singly or combined, may one day enable us to meet and overcome these occasional difficulties. The proper course, nevertheless, appears to be to determine what, in strict propriety, we ought to really regard as tubercle before we proceed to subject it to examination of any kind. Whether what I have advanced will promote this desirable end I do not presume to determine, but would willingly entertain the hope that it may do so.

From the animal chemist, probably, little is to be expected towards the solution of such a question, inasmuch as the albuminous material of tubercle probably differs, chemically, but very little from ordinary albuminous tissues: the modification being, in all probability, rather molecular than atomic; rather a new vital arrangement than a chemical change. The microscope holds out a better prospect of success, and is more likely to recognise and discriminate a peculiar abnormal admixture with natural tissue than any other means or appliances at present at our command.

My friend, Mr. Dalrymple, has kindly undertaken to assist me in this very interesting part of the inquiry; and his great skill and experience in the use of the instrument, and his unimpeachable good faith, afford me grounds to hope that I may ere long be in a condition to lay before the Society results of some interest in regard to minute differences and distinctions, only to be obtained in this way.

TUBERCULAR PHTHISIS.

It may be fairly doubted whether, in perfect strictness of language, the term *tubercular*, taken in an exclusive sense, can with propriety be applied to any form of phthisis; for, however strange and paradoxical it may appear, I venture to submit that, at the present moment, we are not in possession of any conclusive evidence that either tubercle or its disintegration ever constitutes the primary and essential cause of the disorganization which characterises fatal phthisis. We have phthisical disorganization without tubercle, and we have phthisical disorganization with tubercle. In that form of the disease which I have described under the name of tuberculo-pneumonic, tubercles are present; but the tubercles being usually of the sthenic kind, and little compounded, remain, with few exceptions, quiescent, until disturbed by inflammation setting up around them; this inflammation, moreover, proving the chief source of destruction, first of the pulmonary tissue itself, and, consecutively, of the tubercles. Nevertheless, as compounds are, in every instance, more prone to disintegration than simple tubercles, we cannot be surprised that occasionally the most sthenic form of tubercles should, when compounded, undergo primary disintegration; that is to say, disintegration independently of that pulmonic inflammation which constitutes so essential a part of ordinary tuberculo-pneumonic phthisis. It has also been admitted that no precise line of demarcation can be drawn between what, for the sake of distinction, I have called sthenic and asthenic tubercles, the one or the other merely preponderating in different cases. The sthenic, as already observed, preponderate, in the tuberculo-pneumonic phthisis previously described; and if the single term of "tubercular" may be retained at all, to express any particular form of phthisis, I should feel inclined to appropriate it to that sad and fatal form of the disease in which asthenic tubercles prevail, and in which they manifest a strong tendency to assume the compounded character, although it would probably be more correct to employ the term "tubercular" in a specific sense, and regard the two forms of phthisis associated with tubercle as mere varieties: the one, the tuberculo-pneumonic, as the sthenic; and the other, now under consideration, as the asthenic.

This form of phthisis, then, is characterised by a preponderance of asthenic compound tubercles. These often exist in great numbers, either separately, or, more rarely, aggregated together, so as to occupy several adjacent lobules, or even a considerable portion of an entire lobe of a lung. It is not to be understood from this that compound tubercles alone are to be found in this form of phthisis, to the entire exclusion of tubercle in its simple, or even sthenic form: all that is meant is, the great preponderance of compound tubercles. These masses of compound tubercles present various appearances, not only as regards their size, figure, and diffusion, but also according to the stage of development or degeneration in which they happen to be at the period of examination. It would appear, from careful inspection of numerous specimens, that the simple tubercles of which they are made up are sometimes like simple sthenic tubercles, at first extremely minute and of a semi-transparent grey colour, presently, however, assuming more or less of the asthenic character; at other times they are observed to be of an opaque, white, pale-straw or boiled-rice colour, from their very first development, and when of the smallest size. As their size increases, they crowd upon each other, and variously disfigure, as it were, the individual mass or group; whilst some few seem to take the lead, surpassing others in magnitude and rapidity of development, and present greater opacity and more decided loss of cohesion than the rest. These larger, more opaque, and softer tubercles, which project from the surface of the compound mass, on being cut into occasionally, display a manifest central softening; but this is by no means constant: nevertheless, this, and a similar change, now and then observable in the accompanying simple asthenic tubercles, constitute almost the only examples of central softening of tubercle that I have been able to make out.

We thus have the simple tubercles constituting the compound mass, presenting different degrees of advancement; and as the whole proceed to the dull-white, pale-straw, or boiled-rice opacity, thereby increasing contrast of colour, we commonly perceive black stains or streaks variously distributed over them. These have been ascribed to the presence of black pulmonary matter; but although such black pulmonary matter probably exists, it is not difficult to ascertain most distinctly that these black streaks are nothing more than small pulmonic veins containing black blood, a sufficiently distinguishable branch of which may be readily laid hold of by a pair of forceps, and seems to fork or divide, and penetrate the compound mass: so that, however true it may be that simple tubercle is non-vascular, the pulmonic element of compound tubercle continues, at least for a time, to receive blood-vessels.

At a still more advanced period actual softening takes place; sometimes at one point, sometimes at another, and sometimes at several simultaneously or in rapid succession. It is an interesting pathological question, Whence arises this softening? Is it to be regarded as one of the attributes of asthenic tubercle? Does that original deterioration of the albuminous material which unfitted it for organization necessarily lead to the changes which constitute softening, or are we to look for some secondary cause? And, if so, does the secondary cause consist in a process taking place in the pulmonary element of tubercle analogous to inflammation; or is the secondary cause purely mechanical,—the crowding together of non-vascular tubercles mechanically arresting circulation, and thus producing softening of all the tissues and materials of the part analogous to what we see taking place in particular portions of brain, from arrest of circulation through the vessels immediately supplying them with blood? A preparation on the table would rather favour the former of these speculations. It is a portion of the lung of a person who died of phthisis, and whose lungs throughout were thickly beset with tubercles : both the lung itself and the pleura had undergone much inflammation, and it will be seen that to a considerable extent the pleura has been completely dissected from the surface of the lung by the entire destruction of the uniting areolar tissue; whilst simple tubercles remain in great abundance, partly resting on the naked surface of the lung, and partly on the internal, or naturally-attached surface of the pleura. The appearance, too, of the softening mass itself would almost lead to an inference that at all events the softening first takes place in the pulmonary element of compound tubercle; for on attentively examining the softening mass, especially in the most asthenic form of the disease, we can often distinctly perceive the small opaque, whitish or straw-coloured tubercles, as it were, suspended in or supported by a delicate sort of tissue, more or less copiously soaked or infiltrated with a sero-purulent fluid. I may also venture to express it as my belief that the pulmonary element of sthenic compound tubercle not only presents, in many instances, indications of a primary inflammatory process, but even attempts at repair, as shown by hardness, contraction, and blackening of the more central parts. It is, however, of no profit to speculate further on this part of the subject; and whether the softening be the result of a physiological or of a mechanical secondary cause, the fact remains the same;—the compound, and occasionally the simple asthenic tubercles do admit of, or take on, softening, independently of any evidence of inflammation of the pulmonary tissue beyond their respective boundaries.

But although, contrary to the general rule observed in sthenic tubercles, the asthenic variety may, nay, often does, soften, without evidence of either preceding or accompanying inflammation around; it is not thence to be inferred that such primary softening of asthenic tubercle, whether compound or simple, constitutes the sole, or even the principal source of the disorganization of the lung met with in the form of phthisis now under consideration. So far is this from being true, that there, perhaps, never was a single instance of the disease proving fatal, without more or less of this complication. By this complication is not to be understood the inflammation which I suppose to take place in the pulmonary element of tubercle; nor the mere inflammatory state almost necessarily present in the pulmonary tissue, in immediate contact with the softening mass. The inflammation to which I allude is often very considerable: it may extend from the vicinity of the tubercles to one, two, three, or more inches beyond; or it may arise at a distance from the tubercles, having apparently healthy pulmonary tissue interposed; and is moreover very often accompanied by inflammation of the pleura situated immediately above it.

This inflammation usually manifests the same asthenic character observed in the tubercles themselves;—a feeble red hepatization passing at once into disorganization; a grey flabby hepatization undergoing a similar change; complete albuminization of a greater or less portion of pulmonary tissue, presenting the albuminous material in the form of either concrete or fluid pus; and occasionally such rapid exhaustion of the vital properties of the tissue, as to occasion its death, or, in other words, a slough.

As ulceration proceeds in such cases, the albuminous material furnished by the ulcerating tissue, instead of manifesting attempts at repair, by contraction and the formation of a lining membrane or cicatrix, as now and then observed in the more sthenic forms of phthisis, entirely fails to do so, and the destruction proceeds for the most part uninterruptedly till the mischief completely overpowers the constitution and the patient dies. From all this, it may, I think, be legitimately inferred as a general law that *contraction* is, as far as it goes, a favorable sign in every form of phthisis, it indicates a certain degree of constitutional power and attempt at repair. This *contraction*, however, must not be confounded with mere *hardness*, the latter may be occasioned by the mere quantity of tubercular or inflammatory deposit, and it is equally observable in ordinary pneumonic hepatization; but when with hardness we have contraction it may at all times be regarded as favorable.

Accordingly, this contraction is often very inconsiderable in the most asthenic forms of tubercular phthisis, and it is on this account that unless tubercles or inflammatory deposits are very abundant indeed, the physical signs are often much less strongly marked, however extensive the excavation, than in the more sthenic varieties of the complaint, in which, although the excavation may be very limited, the inflammatory contraction and consequent induration of the pulmonary tissue are usually much more extensive as well as more complete. The asthenic character, equally apparent in the tubercles and in the accompanying inflammation, whilst it establishes the influence of constitutional peculiarity goes far to deprive us of all reasonable hope of ever being able successfully to combat this variety of phthisis when once developed, for without repudiating those rare exceptions to be met with in the history of the most fatal disorders, pathology alone would lead me, however reluctantly, to subscribe to the opinion of those who pronounce this form of phthisis at least to be at once incurable and hopeless. *Prevention* I believe to be the golden rule in every form of phthisis, and especially in the tuberculo-pneumonic and tubercular varieties, but I think I may venture to predict that if a remedy, either of prevention or of cure, be in store as a mercy and a blessing for future generations, that remedy will prove to be very different indeed from the enervating expedients so indiscriminately employed at the present day.

I may, in concluding this part of my subject, be permitted to add that the bronchial tubes of the diseased lung very commonly present corresponding indications of want of tone, being thickened, softened, and opaque; and, making allowance for a little exaggeration, may not unaptly, in some instances be compared to a boiled pipe of maccaroni.

In my former as well as in my present communication, I have endeavoured to point out several morbid changes of pulmonary tissue, the result either of recent or of ancient inflammations, which have as I am led to conclude been mistaken for tubercle or tubercular infiltration. I have, at the same time, freely admitted the occasional difficulty of determining positively whether certain changes are attributable to tubercle or to a bygone inflammation, or to a combination of the two. There are, however, some pathological appearances by no means unfrequently met with in phthisical lungs, which, although very commonly regarded as tubercular, are most unequivocally not so.

When a lung has been for some time the seat of irritation of almost any kind, the small bronchial tubes are liable to become hypertrophied—to have their parietes thickened and indurated. On incising such a lung, the truncated extremities of the more minute tubes have very often been mistaken for simple tubercles, in consequence of their canals having been so encroached upon as to escape observation.

The membranous bronchial tubes passing through or near a portion of indurated lung, are very often found considerably and irregularly dilated, so that on examining the lung, the cut surface presents an appearance of several small cavities containing pus, and lined by a pyogenic membrane. These apparent cavities, when of some size and of irregular form, have been regarded as disorganizations, or when smaller and still circular as tubercles softening in the centre. They may, however, by careful examination, be shown to be mere tubes, and a bristle or fine probe may be made to pass beyond the supposed cavity or suppurating tubercle, as well as into the continuation of the canal, found on the corresponding point of the opposite side of the incision.

But perhaps the most frequent source of fallacy in regard to tubercles is to be found in the bronchial tubes which pervade a portion of lung affected with the pneumonic hepatization attendant upon phthisical disease. In this case the parietes of the tube are apt to become considerably thickened, softened, and variously dilated. Such tubes, when they contain puriform mucus, as they generally do, are perpetually being mistaken for tubercles softening in the centre: they may be shown to be tubes by the means already noticed.

An inflammatory albuminous deposit, of some consistency, occasionally takes place in the smaller bronchial tubes, somewhat analogous to what is observed in croup, and in a few cases of bronchitis. This, when the lung is incised, has been mistaken for tubercle.

Neither diagnosis nor treatment forms any necessary part of this communication, nevertheless, at a time when so much is said respecting the facility of recognising and curing consumption, it may not, perhaps, be without its use to append a brief recital of the several forms of thoracic disease which, within my own knowledge and experience, have been pronounced to be phthisis.

1. Recent pneumonic hepatization, especially when situated at the apex of a lung.

2. Recent pneumonic hepatization supervening upon ancient or recent bronchitis.

3. The various forms of pulmonic induration.

4. Pulmonic induration with bronchial irritation.

5. Suppurative, sloughing, or gangrenous pneumonia.

ON THE PATHOLOGY OF PHTHISIS. 63

6. Simple bronchitis, especially when confined to the apex, or otherwise circumscribed.
7. Dilatation of the bronchial tubes with or without induration.
8. General but recent pleuritic effusion.[1]
9. Partial or circumscribed, but recent pleuritic effusion.
10. The flattening of the rib, and compression of the lung, occasioned by ancient pleuritic disease, especially when associated with bronchial irritation.
11. Emphysematous crepitation, especially when coupled with dilated bronchi.
12. Pulmonary apoplexy.
13. Aneurism of the aorta.
14. Malignant disease of the lung or neighbouring parts.

It is well known to many members present that for some years past I have taught it as my conviction that both ancient and recent pleuritic adhesions occasionally give rise, during the respiratory act, to divers modifications of crepitation, varying from that of the smallest subcrepitant to the largest mucous or even gurgling rhonchus. To this opinion, I have made few converts, and for the very obvious reason that the question scarcely admits of direct proof or demonstration. The patient probably continues to live, and if he die, I am unable to obviate the objections urged, that the sounds heard may have resulted either from a mechanical change in the condition or position of the lung, or from some sort of obstruction in the bronchial tubes. Nevertheless, I entertain no doubt whatever about the matter, and, moreover, have much reason to believe

---

[1] Whether the following ought to be regarded as a case of this kind, I do not presume to determine. In the interval between the reading and discussion of this paper, I was requested to visit a young man about twenty-five years of age, whose father told me that he had, the day immediately preceding my visit, taken his son to a gentleman who has excited some attention by the facility with which he professes to recognise and cure consumption. The doctor told him that the case had gone too far; that he must make up his mind to lose his son shortly: that the patient himself needed not to repeat his visit; but that he, the father, might return, and report progress. He was ordered to take a small dose of rectified naphtha twice a day. I found the young man labouring under a good deal of constitutional disturbance, his left chest was full of fluid, and the heart was beating below the right nipple.

My friend, Mr. Cock, at my request, drew off about three pints and a half of fluid, and subsequently about a pint. The patient is doing well.

that the sounds in question have been repeatedly mistaken for phthisical disorganization.

Should this very imperfect communication have the effect of awakening in the breasts of practitioners and students a desire more zealously and carefully to investigate the pathology of phthisis, its principal object will have been accomplished, whether such investigation tend to confirm or to refute the proposition, that *inflammation constitutes the great instrument of destruction in every form of phthisis.*

PLATE IV.

Fig. 1.

Sections of a portion of lung, representing one form of gangrenous or sloughing pneumonia.

a. Pulmonary tissue consolidated by acute pneumonia, undergoing repair, and in transition to permanent grey induration.

b. Portion of lung apparently more in advance towards permanent grey induration.

c. A cavity occupied by sloughing pulmonary tissue.

In this instance the acute pulmonic inflammation proved more severe and extensive than the powers of the constitution could bear. The apices of both lungs had already been excavated by tuberculo-pneumonic phthisis; and hence the inflammation, instead of passing on to contraction and induration, sloughed early, and destroyed life. In this instance there was no fœtor of the breath, which I ascribe to hepatization being complete previous to the slough taking place. In order to have fœtor of the breath, I believe the slough must communicate with a pervious bronchial tube. In this and similar cases the hepatization appears to prevent it.

I have more recently seen a case in which acute pulmonic inflammation supervened upon tuberculated lungs, and in which most extensive sloughing took place early, attended with the most horribly fœtid breath and expectoration. The patient speedily sunk. It was found on inspection, that notwithstanding the extent of the gangrenous disorganization, the inflammation had not in reality passed into perfect hepatization; the pulmonary tissue generally, giving out, on pressure, a considerable quantity of air and serum. Both liver and kidneys were unsound; the two great sources of that cachectic state, so unfavorable alike to the progress and to the results of inflammation; two organs, a diseased condition of which, proves the most insidious and relentless foe both of the physician and the surgeon.

Fig. 2.

This Plate represents a portion of lung affected with recent pneumonia supervening upon simple sthenic tubercles, partial grey induration, and excavation. There is also false membrane on the pleura.

It will be observed that the red hepatization is hardly complete; and is, moreover, of a remarkably livid hue, the usual result of diffuse inflammation in lungs thickly set with tubercles. At the apex we find a very good illustration of the ordinary changes which take place in one form of tuberculo-pneumonic phthisis. The pulmonary tissue in the vicinity of the tubercles has, at some former period, been inflamed; the inflammation has led to iron-grey induration; and portions of this iron-grey induration are now undergoing disintegration, and forming cavities. The ancient pleuritic false membrane, immediately above the indurated parts, sufficiently attests the antecedent inflammation.

a a. The lung beset with simple sthenic tubercles, and passing into a state of red hepatization from acute pneumonia.

b. Pleura thickened by adventitious deposit.

c c. Iron-grey induration from previous pulmonic inflammation around tubercles.

d d. Cavities in the lung, occasioned by disintegration of the old pneumonic deposit, and pulmonary tissue containing it.

e. Bronchial glands enlarged.

Plate IV

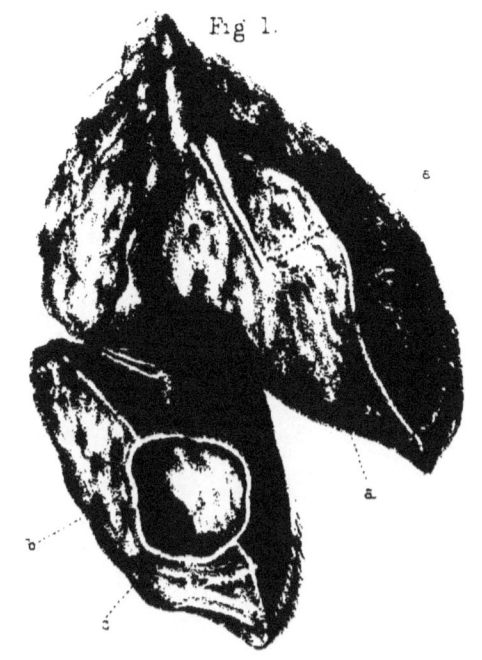

Fig 1.

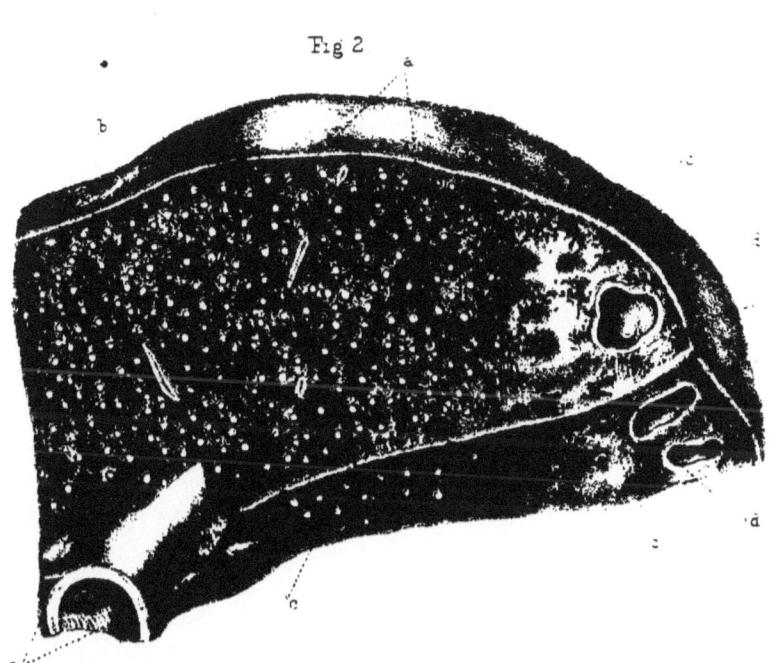

Fig 2

W Hurst del.   W West lith.

PLATE V.

Fig. 1.

A portion of lung affected with red hepatization, and presenting appearances very commonly mistaken for tubercles softening in the centre.

a a. Bronchial tubes thickened, softened, dilated, and containing a muco-purulent fluid. Bristles are introduced into their truncated extremities.

Fig. 2.

Portion of lung divided, to show the opposite sections of an old inflammatory albuminous deposit.

a a. Bronchial tubes thickened and congested.

b b. Apparently healthy pulmonary tissue.

c c. Opake yellow or granular induration, partially surrounded by a portion of lung in a state of iron-grey induration.

d. Portions of yellow granular induration, disintegrated, and occupying a bronchial tube.

e. A bronchial tube closed by the contraction of iron-grey induration.

The little boy from whose lung the drawing was taken, was admitted for recent pneumonia, occurring in connection with the most extensive disease of the dorsal vertebra I ever saw—a connection by no means of very unfrequent occurrence. After death, besides evidence of recent inflammation, the mass of yellow granular induration represented, was found; which, with other appearances, left no doubt as to its having been the result of long antecedent inflammation. There was no appearance of tubercle.

I have more recently seen a very good illustration of this form of pneumonic induration, occurring in an adult female. She had evidently long before, been the subject of pleuro-pneumonia, chiefly attacking the base of the lung. The pleuræ were strongly adherent, both over the ribs and diaphragm; whilst old induration, chiefly of the kind described, was found in detached portions of considerable size at the base of the lung, mixed up, however, with small points of iron-grey, and somewhat more diffused streaks of uniform albuminous indurations, the latter almost resembling merely dense cellular membrane. There were no tubercles.

Fig. 3.

Section of a portion of lung showing red hepatization, partial disorganization, and the truncated extremities of numerous bronchial tubes, with bristles introduced into them.

a a a. Pulmonary tissue in a state of red hepatization.

b b b. Sections of bronchial tubes thickened, softened, and containing a muco-puriform matter.

c c. Cavities from partial disorganization.

Both of the above illustrations were taken from tuberculated lungs; and it may be observed that, unless the surface be naked and uneven, which is not often the case, it is not at all times easy to determine whether the apparent disorganization is not itself, in many instances, a mere bronchial dilatation of unusual size.

Plate V

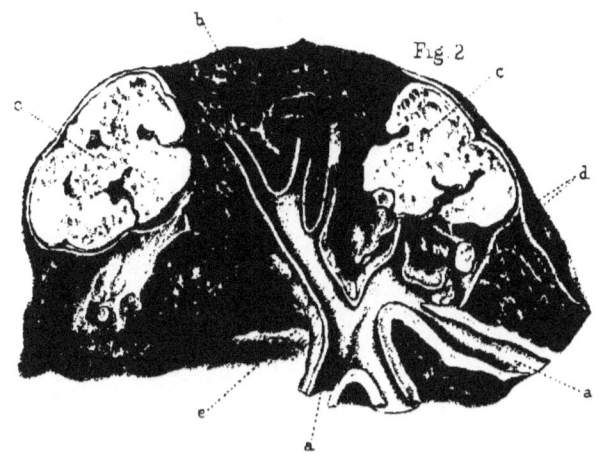

Fig. 2

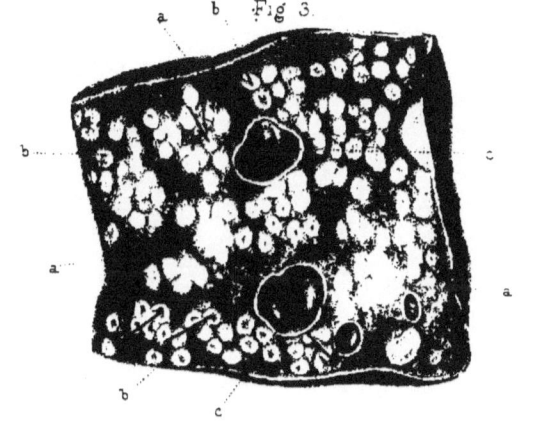

Fig. 3

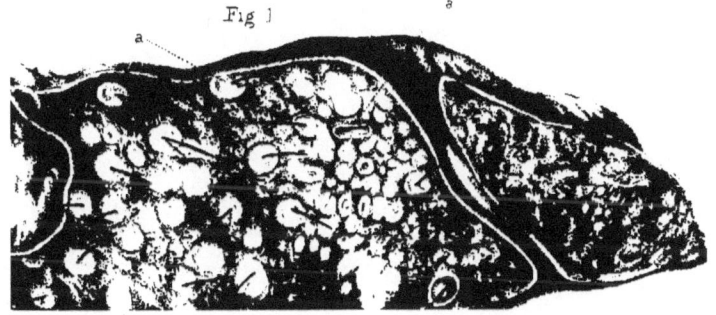

Fig. 1

PLATE VI.

A section of lung, representing tubercles surrounded by the effects of old and recent pulmonic inflammation.

a a. Tubercles surrounded by iron-grey induration from ancient pulmonic inflammation.

b b. Red hepatization from recent pulmonic inflammation.

The apex of the lung had been extensively disorganized by tuberculo-pneumonic phthisis; a form of phthisis in which the disorganization results partly from the softening of recent hepatization, and partly from the disintegration of pneumonic indurations more or less ancient.

Although rude and incorrect, some sort of analogy may be drawn between sthenic tubercles in the lungs and acne on the face. The follicles in the latter, although enlarged and distended, may remain perfectly quiescent, and without producing the least inconvenience; so may tubercle in the pulmonary tissue. Inflammation may be set up in the distended follicle; so may it be, in the pulmonary nidus of tubercle. The inflammation of the follicle may not proceed to suppuration, but terminate in an earlier adhesive process; the same may take place in the pulmonary nidus of tubercle; leading, as it commonly does, to iron-grey induration immediately around. The follicle may suppurate; so may the nidus of tubercle. The inflammation may extend considerably beyond the boundary of the follicle; the same happens with tubercle: nay, in some forms of acne, the intervening skin occasionally inflames primarily; thus, still bearing some analogy to the inflammatory complication, so constantly observed in phthisis.

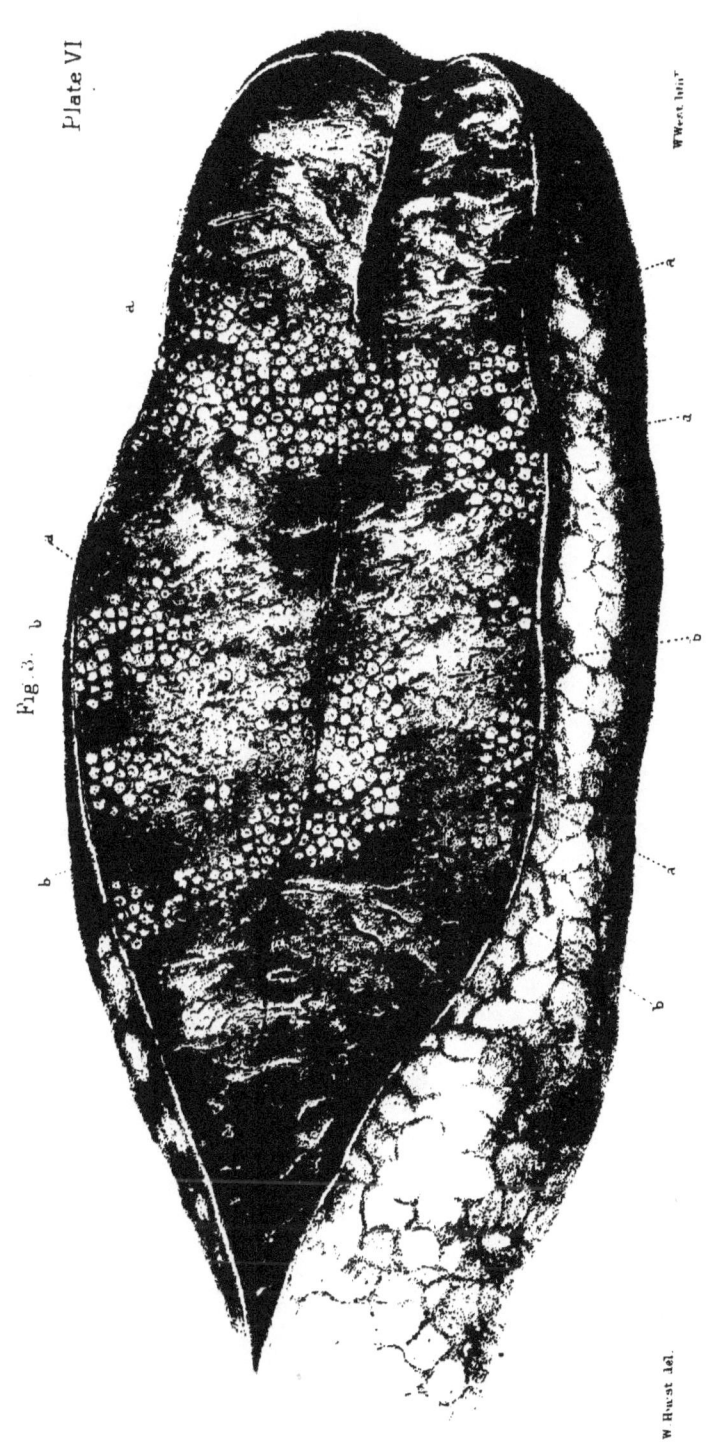

PLATE VII.

The drawing was taken from the lung of a person who died of phthisis, whilst labouring under diabetes mellitus.

The apices of both lungs were more extensively disorganized than I almost ever saw in any form of phthisis: nevertheless, not a vestige of tubercle could be detected in either lung.

It was a case of purely pneumonic phthisis. The case appears to me to be at once curious, and highly interesting, as occurring in connection with the cachectic state observed in advanced diabetes mellitus.

a a. Vast excavations in the apex of the lung.

b b. Smaller excavations in various parts.

c c. Pulmonary tissue in a state of red hepatization.

d d. Pulmonary tissue in a state of grey hepatization.

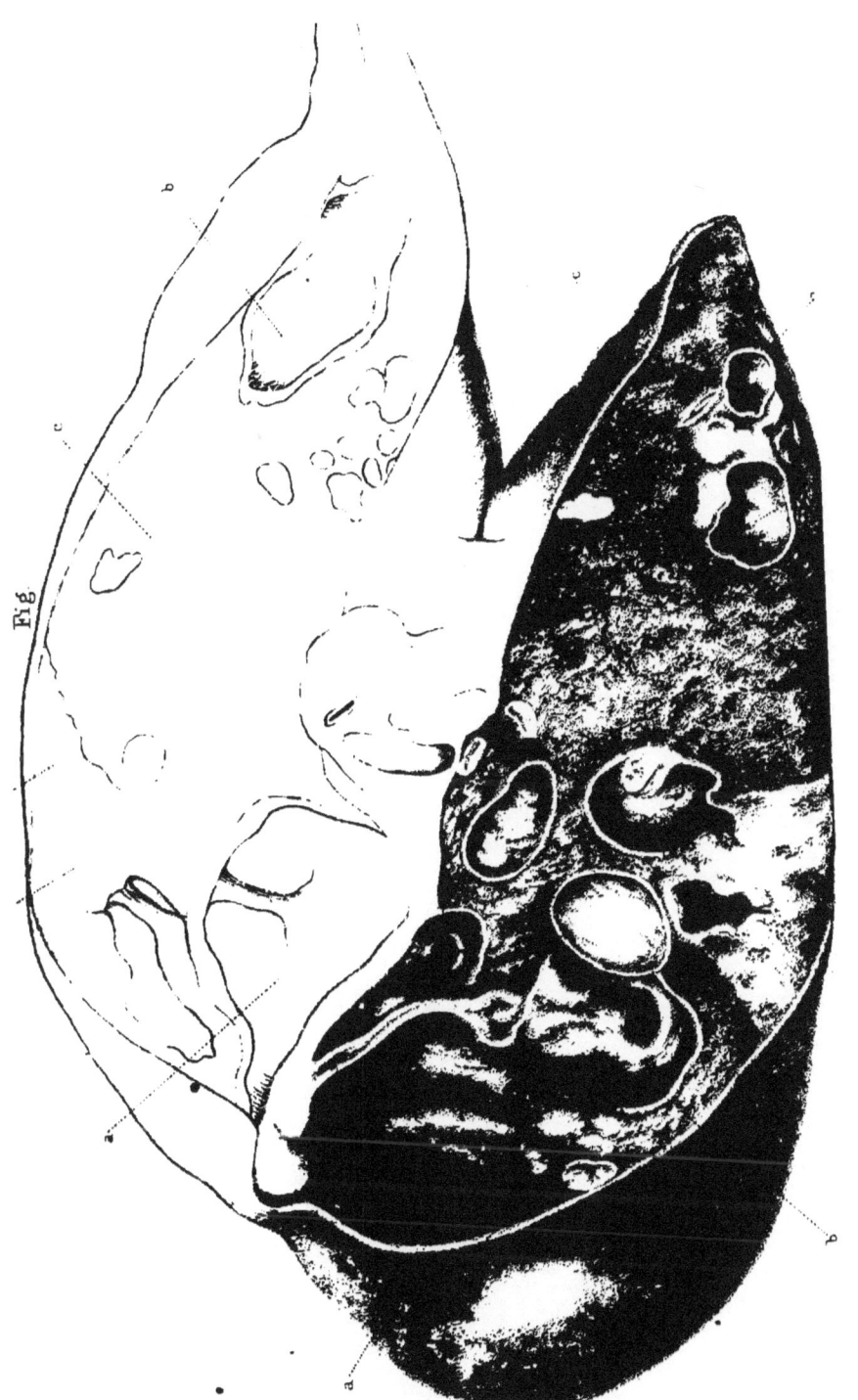

ON

THE DIFFICULTIES AND FALLACIES ATTENDING
PHYSICAL DIAGNOSIS

IN

DISEASES OF THE CHEST.

(*Read before the Guy's Physical Society, February 28th, 1846.*)

WERE I to affirm that Laennec contributed more towards the advancement of the medical art than any other single individual, either of ancient or of modern times, I should probably be advancing a proposition which, in the estimation of many, is neither extravagant nor unjust. His work, 'De l'Auscultation Médiate,' will ever remain a monument of genius, industry, modesty, and truth. It is a work, in perusing which, every succeeding page only tends to increase our admiration of the man, to captivate our attention, and to command our confidence. We are never permitted for a moment to imagine that we are reviewing, for the first time, the mere professions of an ingenious speculator or plausible theorist: we are led insensibly to the bedside of his patients; we are startled by the originality of his system; we can hardly persuade ourselves that any means so simple can accomplish so much, can overcome and reduce to order the chaotic confusion of thoracic pathology; and hesitate not, in the end, to acknowledge our unqualified wonder at the triumphant confirmation of all he professed to accomplish. If to any one this should appear the language rather of exaggeration and enthusiasm than of sober judgment, let him, as a matter of justice to Laennec, endeavour to ascertain the actual state of our knowledge on the subject previous to the time of that distinguished man; let him consult the work of Cullen, the standard work of that day; let him divest himself for the moment, if he can, of all that he has learned directly or indirectly from the labours of Laennec; let him lay aside what he may now regard as the mere alphabet of thoracic disease. I say, let him do this, and he will, I doubt not, be amazed to find that of this very alphabet Laennec was the original and undisputed author.

Great though his merits, incalculable his services to the profession, and indisputable the efficiency of his system of diagnosis, Laennec furnishes no exception to the common lot of all original discoverers: for he, like the rest, had to encounter his full share of doubt and denial, of misrepresentation and ridicule. It cannot, however, be otherwise than gratifying to his admirers and to all lovers of justice to know that he survived his labours and his publication sufficiently long to reap an honourable reward in the respect, the acknowledgments, and cordial congratulations of the most eminent in his profession, and indeed of all who, having had the industry and perseverance fairly to test his merits, possessed a heart capable of rendering honour where honour was due. It is nevertheless scarcely to be doubted, and very much to be lamented, that the cause of Laennec has sustained more real and serious injury from indiscreet and indiscriminate advocacy than from the most determined and open hostility. Credulity may be less unamiable than scepticism: we may be disposed to regard the former as an infirmity untainted by selfishness, whilst we look upon the latter as too often blended with envy, hatred, or malice; still are they alike opposed to the advancement of truth. In the present instance, indeed, the man of science may be permitted to dissent from the moralist; since scepticism, whatever shape it may assume, or from whatever motives it may proceed, serves but to provoke discussion, to encourage inquiry, and to subject all novel pretensions to the searching test of repeated and careful experiment; whereas the tendency of an easy belief is just as certainly to create and strengthen prejudice, to engender misguided enthusiasm, and to pervert or suspend the exercise of that calmer judgment so essential to the honest and impartial investigation of facts.

Books and essays without number, and of great value, have been written for the purpose of adding favourable testimony to the merits of the stethoscope; to increase its utility, and extend its application; whereas, so far as I know, not a single individual has deemed it right or desirable to pursue the opposite course, of expressly publishing to the world the manifold difficulties and fallacies attending its use. The publications alluded to, by the semblance of a too exclusive advocacy, have, according to my humble belief,

placed the stethoscope and its pretensions in a false position; they
have awakened in the minds of many a vague notion of infallibility;
they have led the profession and the public to expect too much,
and by suppressing or concealing the real imperfections of a favorite
expedient, have put it in the power of hostile parties to inculpate
the stethoscope for the errors of the stethoscopist. But if the
works of even able and experienced writers have thus done injury to
the cause of physical diagnosis, and have furnished weapons of
attack to its opponents; what shall we say of that very numerous
class of persons who, with the most slender experience, mistake zeal
for proficiency, and are perpetually falling into grievous and palpable
error? The enthusiasm, the rashness, the bigotry and conceit of
the exclusive stethoscopist have indeed, most seriously retarded the
adoption, and vitiated the claims of physical diagnosis; and have
done more to discourage the student, to shake the confidence of the
profession, and to throw ridicule upon the stethoscope itself than the
most inveterate hostility could ever have accomplished. They seem
to look upon the instrument as all-sufficient; they rush at once to
auscultation and percussion; they neglect or disdain to make those
careful and minute inquiries which no sound and sensible physician
ever fails to do, and thereby convert an invaluable auxiliary into
what, in their hands at least, proves but an imperfect and treacherous
substitute. Unfortunately, in the medical profession, all truly
valuable and practicable knowledge is only to be attained by a proportionate sacrifice of time and labour, and as a general rule the one
may very fairly be measured by the other. Physical diagnosis is
signally obedient to this law; a perfect mastery over it would indeed
be an inestimable acquisition; but its accomplishment can scarcely be
numbered with the possible. The truth is, even moderate proficiency
in the use of the stethoscope is much more rarely achieved than
many are willing to admit; and I venture to affirm that the student
who shall attempt to acquire such proficiency from a perusal of
books, and by an attendance upon patients in the wards of even
this large hospital, will, if he rely solely upon his own individual
efforts, unaided by an experienced guide, most certainly and most
miserably fail in his object. His attempts will prove but a profitless
expenditure of valuable time; he will only be storing his mind with
false knowledge, and in the end will assuredly reap for his reward
disappointment, mortification, and loss of professional fame. The
very language of physical diagnosis must prove, in a great measure,

unintelligible to him; and it would be almost as unreasonable to expect that a man born blind should, on receiving his sight, be able at once to recognise and accurately distinguish the ever-varying tints of a landscape, as to suppose that a novice in the art of physical diagnosis could, without the aid of an interpreter, uniformly attach a correct idea to every term employed to express the ever-varying phenomena elicited by auscultation and percussion. It may indeed be alleged that the example of Laennec himself presents us with unquestionable proof how much may be accomplished by the unaided exertions of an individual. This is undoubtedly true; but the fine genius, the indefatigable industry and vast opportunities of Laennec do not fall to the lot of many, and his case can only be regarded as a very rare exception to a very general rule. Laennec not only gave assistance, but a language to the art. Scarcely more hopeful is the case of the practitioner who, having unfortunately neglected the cultivation of physical diagnosis during his pupilage, has already entered upon the duties of his profession. Driven by necessity or pride to rely upon his own resources, he will, even with all the advantages of a large hospital, seldom be successful; and if without the benefit of such opportunities as a large hospital affords, he would probably be exercising a sound discretion were he to repudiate the practice altogether, or, at most, make it entirely subservient to other means of diagnosis.

Such sentiments, proceeding from a hospital physician, may by some be pronounced to be vain and presumptuous. Nevertheless, my conscience tells me that in the present instance, plain speaking, if not the most prudent, is certainly the most honest policy. To strip the stethoscope of the extravagant and meretricious pretensions thrust upon it by injudicious friends; to make a candid acknowledgment of the manifold difficulties and fallacies to be encountered in its employment; to state fairly what it will not, as well as what it will do, is surely calculated to render it some service: for by so doing we disarm hostility and establish on a solid foundation its legitimate claims to the respect and confidence of the profession. If, notwithstanding this explanation, my sentiments shall be deemed an offence, it is to be hoped that the confessions, which are to follow will go some way towards its excuse: neither will it be taxing generosity too much to claim an acquittal from all empirical motives, if I venture to declare, what is well known to many members of the Society, that these confessions are made by one who has zealously cultivated the practice of auscultation and percussion at this large hospital and elsewhere for a quarter of a century.

Desirous on the present, as on all other occasions, of economizing the time of the Society, and promoting discussion, I have arranged my subject in the form of separate propositions, and have attached a number to each. By such an arrangement, any single proposition can readily be singled out and commented upon by those gentlemen whose intention it is to favour us by joining in the discussion.

PROP. 1.—*It is well known that many persons whilst under examination entirely fail to perform the respiratory act efficiently, either from mere nervousness or from a mistake in regard to the mode of accomplishing it; they merely heave the chest up and down instead of freely and forcibly inhaling and expelling the air. This may lead to an erroneous belief that the respiratory murmur is deficient, even whilst the lungs are perfectly healthy.*

This source of fallacy is best corrected by desiring the patient to cough, and instantly afterwards to inspire forcibly in order to cough a second time. When to this expedient is added a comparison of the two sides of the chest, the actual condition of either lung may, in general, be determined with considerable precision.

2. *Whatever lessens the freedom, mobility, or elasticity of the ribs, renders the sound, on percussion, more dull. Hence it is that in rickety persons, where deformity of the chest has taken place subsequent to birth, the signs furnished by percussion are often extremely unsatisfactory; and, indeed, under such circumstances, neither percussion, nor, in many instances, auscultation, can be much relied upon.*

3. *Some persons, without actual deformity, have naturally such fixedness of the ribs, that they at all times manifest very imperfect resonance, as well as considerable feebleness of the respiratory murmur.*

4. *The rigidity of the cartilages of the ribs in advanced life has a similar effect, and, moreover, often tends to throw obscurity over hypertrophy of the heart, by preventing the usual heaving of the ribs at each systole of the hypotrophied organ.*

"The dulness of the sound elicited by percussing the chests of aged and deformed persons arises probably from the fixity of the ribs at their extremities. In aged persons this arises from ossification of the cartilages and rigidity of the tissues; and in deformity from a permanently unnatural elevation or depression of the ribs. In such conditions the force of a sudden blow upon a rib is transmitted to the parts at its extremity, and this occurs the more readily from the convexity of the parietes. Whereas, in the condition of health, such a force applied to a rib would be in a great measure limited to it, the yielding attachments of either end allowing of its vibrating. Looseness of connection or fixation of the extremities will produce much greater difference in the cone of a curved rod, the force being applied to the convexity, than when the rod is straight. In the former, the force coming upon the keystone of the arch is transmitted on either side to the extremities of the curve; whereas, in a straight rod, the force applied curves it from the force, and tends to loosen the connection at the extremities, its own elasticity causing it to return to its previous condition, and so to vibrate, although the ends be fixed."[1]

5. *When exploring the chest in a case of recent disease, we may be misled by the permanent effect of an ancient pleurisy.*

Such a sequel to pleurisy may, though rarely, involve both sides of the chest, and may be so considerable as to resemble externally the deformity which usually results from rickets. More commonly we find the deformity of old pleuritic contraction at the inferior part of the chest only, either anteriorly or posteriorly, or both.

This deformity, from pleuritic contraction is sometimes obvious to the sight, but even when not so apparent, the difference of the two sides, when one only is affected, may in general be recognised by simply grasping from before to behind the ribs of one side with the right, and the ribs of the other with the left hand, and comparing the respective rotundity or flattening of the two. The physical signs also of ancient pleuritic contraction, although often obscured by the more recent disease, are in general pretty decisive; these are dullness of sound on percussion, feebleness or absence of respiratory murmur, constrained movement of the ribs, the dry crepitations or cracklings, especially during inspiration, the harsh dry croaking sound or *rale sonore* during inspiration, to which may be added, the incom-

[1] Dr. Gull.

patibility of a few, or of many of these with the recent history of the case, and with other physical signs present. In such a case I consider the employment of the naked ear much superior to that of the stethoscope. When ancient pleuritic mischief exists at the upper part of the chest, and especially in front, it occasionally proves a source of very serious mistakes, to be noticed hereafter.

6. *When, as usually happens, rickety deformity consists in lateral flattening of the ribs, with projection of the sternum, the action of the heart is liable from slight causes to beat with such violence, and to have its sound and impulse so extensively diffused, as not unfrequently to have led to an unfounded apprehension of serious organic disease of that organ.*

I had an opportunity of witnessing a good illustration of this some time ago, with my friend Dr. Ridge. The ready disturbance of the heart's action, and the great extent of its sound and impulse had led to a belief that it was organically diseased, yet inspection after death displayed a healthy organ.

7. *When acute disease of the lungs occurs in persons with rickety deformity, the violence of the symptoms is often quite disproportionate to the extent and severity of the pulmonary disease, and may thereby suggest unnecessarily active treatment.*

8. *When the abdomen is greatly distended with fluid, the encroachment of the diaphragm upon the chest and its imperfect descent during inspiration, often give rise to such dulness on percussion, and feebleness of respiratory murmur at the inferior part of the chest, as may be mistaken for effusion into the latter cavity.* On the right side enlargement of the liver, and on the left enlargement of the spleen, may to a certain extent, have a corresponding effect.

9. *Of all the sources of fallacy to be encountered in the physical diagnosis of diseases of the lungs, bronchitis is by far the most prolific of mistakes and oversights. It may greatly obscure phthisis, pneumonia, and pleurisy; as well as divers forms of chronic disease of the organs.*

10. *When the bronchitic complication in phthisis is considerable, we often fail to detect some, or all, of the ordinary physical signs of*

the latter—dulness on percussion, tubular respiration, and even bronchophony and pectoriloquy. This is more especially the case, however, in the earlier stages of phthisis; the difficulty being then increased by the absence of any flattening, or even immobility, of the ribs of the side affected.

It is under such circumstances that the too exclusive stethoscopist is liable to be beaten in diagnosis by those who reject physical examination altogether, for the latter inquiries carefully into the history of the patient and of his family, he observes attentively the patient's general aspect, and the character and order of the general symptoms; all of which are wont to be too much disregarded by the stethoscopist. Nevertheless, if the stethoscopist do his duty, he has greatly the advantage, he will institute an equally careful inquiry: added to which, he will observe whether the bronchial obstruction is limited to the apex of the lung, he will repeatedly and for some minutes at a time, apply the stethoscope, or what I prefer, the naked ear, to the upper part of the chest, he will desire the patient to breathe freely, to cough, and if possible to expectorate. He will, by so doing, often succeed in removing the obstructing mucus, and thereby develope, however slightly, some degree of bronchial respiration, or bronchophony, or both, signs strongly confirmatory of phthisical disease in doubtful cases. According to my own experience the individual symptom which, without being decisive, above all others increases the apprehension of phthisical disease, is occasional slight hæmoptysis just sufficient to tinge or streak the sputa. Little reliance can, I fear, be placed upon any other appearances of the sputa, as distinctive of the two diseases. The more limited the bronchitis, especially if situated at the apex, the greater the probability of its being associated with phthisis; the more general the bronchitis, especially if it affect both lungs, the greater the hope of an exemption from phthisis. The more abrupt the transition from bronchial obstruction to natural or puerile respiratory murmur in the affected part of the lung, the greater the likelihood of phthisis.

Moreover, when there are no indications of bronchitis of the small tubes, and its almost constant associate, vesicular emphysema; constituting a very common form of what is vulgarly called asthma, a puerile or blowing murmur, wherever situated, ought at all times to create a suspicion of phthisical or other form of consolidation in the adjacent pulmonary tissue; although neither auscultation nor percussion may detect its presence at the time. On the other hand, in the form of bronchitis alluded to, such puerile or blowing murmurs are so very common, that whenever the long, laboured, wheezing expiratory murmur, or paroxysmal character of the dyspnœa indicates the former, we may at all times expect to find more or less of the latter.

The general tendency of bronchitis is to enlarge, that of phthisis to contract the chest.

11. *When with bronchitic rales the stethoscopist detects some dullness of sound on percussion, tubular respiration, bronchophony, pectoriloquy, and gurgling; it still is not conclusive evidence of phthisis, as the whole of these signs may result from the permanent changes produced from a former pleurisy, pleuro-pneumonia, or whooping-cough, or even a recent pleurisy or pneumonia, when these several conditions happen to be associated with considerable bronchitis.*

There is a case in Lazarus Ward, under the care of Dr. Barlow, somewhat illustrative of this proposition. We have the history of ancient pleurisy, there is universal dulness over the right side, with contraction of the ribs, vibration is not extinct, the patient expectorates puriform matter, and there are all the signs of excavation in the lung. The case is, in every respect so strongly marked that no one can doubt its original character, although before death I hold it to be impossible to pronounce with certainty the existence or absence of a cavity.

The case of Robert B—, aged 28, admitted into our Clinical Ward, resembled the above, in several particulars. He had been in the hospital four years before with the sequelæ of pleurisy—general dulness over the whole of the right side of the chest, contraction of the ribs, heart drawn over to the right side, all the ordinary signs of an excavated lung, and copious puriform expectoration. When he reappeared in our Clinical Ward, with the exception of greater emaciation, his general condition and the state of the right chest were little altered.

Acute disease set up in the opposite lung destroyed life, and dissection proved the correctness of the opinion that had been formed as to the pleuritic origin of the disease in the right side. The remarkable displacement of the heart excited some surprise, nevertheless, I am disposed to conclude, from what I have seen in practice, that great contraction of the right chest after pleurisy almost as certainly draws the heart towards the same, as extensive effusion into the left chest forces it towards the opposite side.

When the changes just described are more limited in extent; and more especially when they occur at the upper part of the apex of the lung, a correct diagnosis becomes much more difficult. The compressed or condensed lung, the dilated tubes, and copious bronchial secretion give rise to signs perfectly identical with those of disorganising phthisis, and although the former may prove fatal, either by gradual emaciation and exhaustion, or by setting up actual ulceration in the pulmonary tissue, still our prognosis would be very different in the two cases—the one holding out a fair and reasonable hope of recovery, whilst the other would be regarded as nearly or altogether hopeless.

The history and progress of the case, together with a careful consideration of the symptoms present and past, will afford us the best grounds on which to found our diagnosis.

12. *When, in phthisis, the larynx becomes so involved as to impede the entrance of air, and thereby give rise to a permanent rale sonore in that organ, the reverberation of the rale through the entire chest, at each inspiration, greatly obscures the stethoscopic signs, and often leads to a mistaken belief that the obstruction and imperfect respiratory murmur are attributable to the lungs themselves, when the latter, though perhaps unhealthy, are in a condition totally different from that suspected.*

This source of fallacy is detected by applying the stethoscope to the larynx itself, and by exploring both sides of the chest, when it will be found that the *rale sonore* is audible everywhere.

13. *When, with partial obstruction in the larynx, there is complete loss of voice, the results of mere auscultation are often of very little avail in diagnosis.*

14. *When the bronchi opening into a phthisical cavity are temporarily obstructed by secretion, auscultation may fail to detect that cavity, especially if the patient breathe but moderately, and should the cavity be large and superficial, the fallacy may be rendered more complete by a certain degree of resonance being elicited by percussion.*

In every case of suspected phthisis, therefore, we ought to cause the patient to breathe and cough with some violence, and repeat the experiment from time to time, whilst the ear continues to be applied to the chest. I have known large cavities overlooked from a neglect of these precautions: the puerile respiration, which so often surrounds phthisical obstruction, tending not a little to promote the fallacy.

15. *A person may have a violent tearing cough, lasting for weeks or months, attended with slight mucous expectoration, occasionally even streaked with blood, and causing pain to be felt through the whole chest, as well as an appearance of general constitutional distress, whilst neither auscultation nor percussion can detect any morbid change within the chest.*

All this often results from a relaxed uvula, in either sex; or, if the patient be a female, it may depend upon hysterical irritation. More rarely I believe it to occur in connection with miliary tubercles antecedent to phthisical disorganization.

16. *When any form of chronic induration of the pulmonary tissue exists, and especially if attended with dilated bronchial tubes, neither auscultation nor percussion enables us to distinguish such a condition of lung from phthisical disease. If bronchitis be present, and the induration be situated at the apex, the signs are perfectly identical with those of phthisical disorganization.*

Our diagnostic resources are to be found in the history and progress of the case, the absence of some of the more ordinary symptoms of phthisis, and some incongruity observable between the local signs and the general aspect and condition of the patient. Illustrations of this proposition are of repeated occurrence.

17. *Auscultation and percussion alone are insufficient to distinguish malignant disease, hydatids, or a tumour from more ordinary diseases of the chest.*

Nevertheless, by disclosing something unusual, or apparently incompatible with ordinary diseases of the chest, physical examination will often excite a suspicion which—by a careful consideration of the history and progress of the case by the wooden dulness of sound on percussion, by the peculiar expectoration, by the effects of mechanical obstruction on the veins of the neck or external chest, or even by exploration with a fine needle or trochar—may occasionally be converted into certainty.

A very remarkable fallacy of the opposite kind came recently under my own observation. A lady whom I had attended professionally for several years had occasion to travel whilst in a very delicate state of health. After much suffering from increasing illness, she at length, in returning, reached Paris, and there consulted one of the highest medical authorities. Finding extreme dulness in the upper part of the chest, and some resonance remaining below and behind, the signs and suffering were pronounced to arise from some solid growth. As I had examined the chest a short time before her departure, as I knew that she had diseased heart after rheumatism, and that a former attack of pleuro-pneumonia in all probability had left adhesions, I could not but conclude that fluid effusion was the cause of her distress. I discharged several pints.

18. *If acute pneumonia have already proceeded to complete hepatization when we first examine the patient, the physical signs are not unfrequently insufficient to distinguish the morbid change from phthisical disease, or from ancient pulmonic induration, with or without dilated bronchial tubes. This is more especially the case when acute pneumonia assails the apex of a lung, which is by no means very uncommon.*

I have on several occasions known hepatization of the apex from acute pneumonia, pronounced to be phthisis by stethoscopists; they have not sufficiently appreciated the difficulty, they have neglected to inquire carefully into the history and progress of the case, and have mistaken the pungent heat of skin of ordinary pneumonia for that which occurs in phthisis, and which I believe, nevertheless, has often the same origin.

"Mary B—, aged 19, admitted, under Dr. Addison. She had always been delicate; and after the whooping-cough and measles, which she had had eight years before, had been subject to attacks of cough and cold, in which she had frequently expectorated blood.

"When admitted, she had a troublesome cough, with scanty sputa, slightly tinged with blood. There was dulness on percussion below the left clavicle, with tubular breathing and gurgling, the last extending down nearly to the margin of the ribs, where it became more dry and crepitating. Posteriorly there was dulness on the left side, extending from the apex nearly as low as the angle of the scapula; and tubular breathing and bronchophony to the same extent; slight tubular breathing and bronchophony at the right apex, with large crepitation.

"Under the impression that phthisis was present, she was ordered,

> Empl. Canth. infra claviculas singulas applicetur.
> Pil. Papav. c̄. Ipecac. bis die sumat.
> Mist. Mucilag. ter die sumat.

Under this treatment the cough became worse; sputa more copious, fawn coloured, and uniform, except that it contained puriform streaks; the head painful; pulse 120, compressible; cheeks flushed; skin hot and pungent; while the signs afforded by auscultation and percussion continued unaltered. She was then bled twice to eight ounces; and calomel, antimony, and opium were administered. Her mouth was kept sore by the calomel for a few days, when it was discontinued, and nothing given but julep ammon. acet. Under these remedies, the dulness, tubular breathing, bronchophony, and gurgling at the left apex, gradually diminished, and at length entirely disappeared. The tubular breathing and bronchophony at the right persisted longer, and caused some alarm; but these also ceased: the expectoration became white and frothy, and then, with the cough, subsided.

19. *When pneumonia occurs in its simplest form—that is, with little or no bronchial complication—there is sometimes no cough, and consequently no expectoration; the whole case so closely resembling common continued fever, that both the stethoscopist and the non-stethoscopist are apt to be thrown off their guard.*

In such a case the stethoscopist has greatly the advantage; for if he do his duty he will pretty certainly detect the pneumonia by physical examination; whereas the non-stethoscopist is very likely to remain in ignorance of its existence, to the serious detriment or destruction of the patient.

Although there are many exceptions, this, the simplest form of pneumonia, most frequently occurs in the aged and cachectic; and in such subjects I have met with a few cases in which even an ordinary physical examination of the chest might have failed to detect the disease. A case of the kind occurred some time since in Billet Ward, and to which I remember having directed the attention of Dr. Gull. On desiring the patient to breathe, neither crepitation nor tubular respiration could be heard; but on urging him to inspire violently and to cough, both signs became sufficiently developed to declare the presence of the disorder. It would appear as if the muscular strength were so much impaired, that, unless by an unusual effort, a sufficient expansion of lung does not take place to admit the necessary impulse being given by the air to the tubes and cells of the organ.

It is in such cases that the very characteristic pungent heat of skin so often and so fortunately directs attention to the chest. This pungent heat, though not necessarily present, is very rarely absent so long as any portion of lung continues to be in the first or crepitating stage of pneumonia.

20. *Physical examination of the chest does not enable us to distinguish an advanced stage of pneumonia with considerable bronchitis from pneumonia with breaking up of the lung, a difficulty the more embarrassing inasmuch as the former may pass into the latter.*

Here we must be guided chiefly by the history and general symptoms. The phthisical aspect, the cachectic or exhausted condition, or the intemperate habits of the individual will afford us some help in arriving at a correct diagnosis.

21. *When the anterior and inferior portion of the left lung is consolidated by pneumonia, percussion may produce good resonance, in consequence of the proximity of the flatulent stomach, and thereby throw us off our guard. When pneumonic consolidation takes place anteriorly and inferiorly, and even posteriorly, on the right side; a remarkable degree of resonance is occasionally elicited in a highly tympanitic condition of the intestines. Under precisely similar circumstances auscultation may detect a well-marked modification of amphoric respiration, and metallic tinkling, to a considerable height in the chest, thereby leading to the erroneous conclusion that pneumothorax is present.*

The respiration acquires its amphoric character by reverberating through the solid parts of the inflated bowels, the metallic tinkling developed below the diaphragm acquires its intensity by reverberating in the opposite direction.

The following case presents a very remarkable modification of this source of fallacy:

"Charlotte C—, aged 19, was admitted into our Summer Clinical Wards, complaining chiefly of some pain and tenderness over the whole of the abdomen, but especially on the right side, and the bowels appeared to be greatly distended with flatus. These complaints were of a month's standing, but about three months prior to admission she had had an attack of inflammation within the chest, and was still harassed by a short dry cough.

"On the right side of the chest, anteriorly, there was increase of resonance as high as the third rib, and even the apex, the resonance was greater than normal, dulness at the base of the same lung posteriorly.

"Respiration was puerile in the apices of both lungs anteriorly, mixed with sibilant rales in the right. On the right side, as high as the third rib, slight and distant crepitation, with a metallic state of the breathing, and when she spoke or coughed similar amphoric sounds were to be heard. Accompanying these amphoric sounds we had the *tintemint metallique*."

On inspection after death, the pleuræ on both sides of the chest were adherent; but the diaphragm had been raised up high within the chest, partly by the inflated bowels, and partly on the right side, by old adhesions between it and the base of the lung. A vast fæcal abscess extended from the pelvis to the under surface of the diaphragm on the right side. From this inflated abscess, or from the distended intestines, or from both, had originated the great resonance, the amphoric sound, and *tintement metallique*.

22. *Physical examination cannot determine whether pneumonia, in any of its forms have or have not supervened upon tubercles, although the prognosis in the two cases would be very different.*

23. *I very much doubt whether physical examination can in any instance determine with certainty the existence of simple tubercles in the lungs.*

In a former paper I communicated a case which very well illustrated this, and the following is to the same effect:

"Adah H—, aged 17, admitted under Dr. Addison. She was

admitted with what appeared to be symptoms of fever. The chest was resonant on percussion, and the respiration puerile. It was supposed that the breathing was somewhat harsher at the left apex than at the right. There was no cough or expectoration. On the following day, the case proved to be one of hydrocephalus, and on the next day she died. At the post-mortem examination, numerous miliary tubercles were found scattered throughout the lungs: they were more numerous in the upper and middle lobes than in the inferior; and in the right lung, than in the left, the latter being the lung in which the respiration was considered harsher. The pulmonary tissue surrounding the tubercles was perfectly healthy.

24. *We may be called to a case of pleurisy before a single physical sign has been developed.*

In such a case it may be doubtful whether the pain arises from pleurisy, rheumatism, neuralgia, or the approach of shingles.

A very good example of the latter source of fallacy presented itself in the case of Sarah F—, aged 18, admitted into the Summer Clinical Wards. It will be found accurately described in the clinical book by my clerk, Mr. Greenwood.[1]

25. *When pleurisy occurs low down in the angle between the ribs and diaphragm, and especially when situated anteriorly, a considerable period, perhaps several days, may elapse before auscultation can detect either pleuritic rubbing, ægophony, bronchophony, or tubular respiration, whilst percussion proves fallacious, in consequence of the presence of the liver on the right, and of the inflated stomach on the left side.*

Doubtless this source of error is connected with the relative positions of the effusion, the lung and the parietes of the chest, but without attempting any elaborate explanation, I content myself by simply vouching for the fact, in my own experience. I have known such cases mistaken for spasms, for neuralgia, for hepatitis, for splenitis, for peritonitis, and, in consequence of pressure of the abdomen causing a descent of the false ribs, for enteritis. A knowledge of

[1] The Summer Clinical Wards were committed to my charge, and were appropriated exclusively to diseases of the chest. Dr. Novelli, Mr. Greenwood, Mr. Walter Johnson, and Mr. Howard Johnson, were my clinical clerks; and better I could not have had.

the fact alone will go far to preserve us from error, and when the suspicion exists a long-continued application of the naked ear to the chest will often succeed in detecting a little transient graze, or pleuritic crepitation, sufficient to greatly strengthen that suspicion. The lapse of one, two, or three days will in general place the matter beyond dispute.

"Jeremiah —, aged 23, ill two weeks; was admitted into Job Ward under Dr. Addison. He complained of acute pain beneath the right nipple when a deep inspiration was taken, of headache, pains in the limbs, and extreme prostration. There was a frequent, harassing cough, copious expectoration of a frothy mucus, tinged with blood, the skin was pungently hot, and the tongue was loaded with a moist white fur.

"*Physical examination of the chest.*—It was deep and well formed. The movement of the right side was instinctively checked during respiration. Vocal fremitus good in all parts. From a peculiarity in the configuration of the thoracic parietes, not very unfrequently met with, percussion elicited in all parts very imperfect resonance. Dulness, it is true, was more strongly marked beneath the right mamma, and below the scapula of the same side, than in the corresponding parts of the left lung, but the presence of the liver renders this sign fallacious, and but imperfectly diagnostic. The signs afforded by stethoscopic observation were, if possible, still more unsatisfactory. Where the sounds of respiration could be fully heard, with the exception of occasional bronchitic whines, they were normal. At the lower parts of both lungs the patient will not permit the air fully to enter the cells, and even after coughing, nothing, save a slight roughening of the respiratory sounds before, and a tendency to tubularity behind in the right lung could be detected. The voice was triflingly modified at the base of the right lung: it was bronchophonic in character.

"The case was diagnosed to be phlegmasial fever, with bronchopleuro-pneumonic complication, and Dr. Addison ventured to prophesy that in a few days, all the stethoscopic signs diagnostic of such pulmonic disease would present themselves.

"The treatment consisted in moderate depletion, local and general, with small doses of antimony, opium, and calomel combined in form of pill.

"On the second and third days no change took place in the physical signs, save that on the latter, after long listening, a soft rubbing sound attendant on respiration was occasionally heard in the angle below the right mamma.

"On the fourth day, a distinct to and fro sound, accompanied by moist crepitation, had become audible in the right submammary region. The friction sound was most marked to the right of and on a level with the nipple. It diminished in intensity as the angle was approached, where it was replaced by fine crepitation. At the base of the lung behind, some minute crepitation had arisen, and immediately below the scapula a soft brushing to and fro sound could occasionally be heard. This examination was made at 11 a.m. In the evening of the same day the patient was re-examined.

"In front, little change had taken place excepting an increase in degree of the previous signs. The friction sound was more harsh, and the crepitation had become larger. Behind, harsh grating to and fro sounds—with which was mingled crepitation in various degrees, large and small, extending from the base of the lung over the lower half of the scapula and round to the axilla—had become audible. Diffused rhonchi were to be heard in various parts. All the signs, in short, that are recognised as diagnostic of bronchitis and pleuro-pneumonia had manifested themselves."

The further progress of the case is detailed in the clinical books.

"Emma H—, aged 19, a tolerably healthy-looking girl, was admitted into the Clinical Ward, under the care of Dr. Addison, Nov. 26, 1845. She had been labouring for several months under some obscure affection about the pelvis, which, after careful examination, appeared to be probably nothing more than hysterical. Three weeks after her admission she was attacked with a severe rigour, followed by great heat and dryness of skin; these symptoms were succeeded on the following day by an acute pain in the right side of the chest, below the mamma, increased by inspiration, and attended with a short hacking cough and all the ordinary symptoms of pyrexia. The chest was attentively examined by Dr. Addison, myself, and others, but none whatever of the physical signs of pleurisy could be detected.

The next day she had great anxiety of countenance, distress of breathing, and, in short, an aggravation of all the symptoms. On listening for some time towards the base of the right lung posteriorly I thought that a pleuritic rub or friction sound was occasionally audible at the end of a forced inspiration, but Dr. Addison after a protracted examination said he could not hear it.

"The following day, viz., the third from the attack, there was no doubt as to its existence, it having become constant, although heard over a limited space, and unaccompanied by the signs of serous effusion."

Some time ago I attended a very instructive case of this kind, with Mr. Busk and Mr. Bell. It occurred in the person of a foreigner. He was supposed to be suffering from acute hepatitis. The chest had been carefully examined, and no physical signs of disease within that cavity could be detected, the dulness present on percussion being attributed to the liver. I repeated the examination, but although persuaded that it was a case of pleurisy affecting the angle, and from the pungent heat of the skin that the pleurisy was associated with pneumonia, I could detect no sign whatever, except some dulness on percussion, and there was neither cough nor expectoration. A few days removed all doubt about the matter.

I may observe, in concluding my remarks on this proposition, that when pleurisy has its seat in the parts alluded to above, it constitutes by far the most painful and perhaps the most dangerous form of the acute and sthenic disease. It is the paraphrenitis of the ancients, a disease which, according to them, consisted simply of inflammation of the diaphragm. This, however, is not correct, for the pleura covering the diaphragm is often inflamed without giving rise to the dreadful suffering observed in paraphrenitis, whereas, when acute and sthenic inflammation attacks the pleura, where it is reflected from the diaphragm of the ribs at the base of the chest, and thus involves both the diaphragmatic and costal pleuræ at the same time; then it is that we have such intense suffering and such an expression of agony in the countenance, as forcibly to remind us of the *risus sardonicus* of the older writers.

26. *When the effusion into the chest is of the purely serous kind, or when the proportion of albuminous material is very inconsiderable, the fluid gravitates to the floor of the cavity, and may, unless very abundant, entirely escape detection, either by auscultation or percussion.*

The lung rests, as it were, on the surface of the fluid; it still admits air, and consequently general expansion; it still continues to apply itself more extensively to the parietes, and consequently neither ægophony, bronchophony, nor tubular respiration is necessarily present in a marked degree, whilst any slight dulness of sound on percussion is rendered equivocal by the liver on the right and by the flatulent stomach on the left side. Neither does vocal vibration necessarily cease under such circumstances. The correctness of this proposition is not, I am aware, generally admitted, any unexpected scrous effusion being under such circumstances supposed to have taken place immediately before dissolution, or to be purely cadaveric. Experience, however, leaves me no room for doubt about the matter. In such cases we must found our opinion, or rather our suspicion, upon the history and general condition of the patient. We know that effusions of the character described are most likely to take place in diseases of the heart, in cachectic and dropsical persons, in those suffering under Bright's disease, or the sequelæ of scarlet fever. I may, nevertheless, be permitted to observe that when there has been a suspicion of the presence of serous fluid without any well-marked ægophony, bronchophony, tubular respiration, or unequivocal dulness of sound on percussion, by placing the patient in a sitting position, and placing the ear for a considerable time to the posterior and inferior part of the chest, and in the mean time moving the patient's body several times to nearly the horizontal position and back again, and directing him to cough forcibly, I have succeeded in eliciting a distant tubular respiration, sufficient to indicate partial compression of the lung.

In the case of John B—, aged 49, affected with Bright's disease, the signs of effusion into the right chest were very unsatisfactory, although on a post-mortem examination shortly afterwards, it was found to contain a large quantity of serum.

Frederick S—, aged 24, laboured under disease of the heart of long standing, but greatly aggravated by bronchial complication. Bronchial râles were loud throughout the chest, the right side was generally dull on percussion, there were some bronchophony and very imperfect respiratory murmur, the whole of these signs being most strongly marked posteriorly and inferiorly. *Vibration was nearly perfect.*

A difference of opinion arose whether the signs depended upon old pleuritic adhesions or upon serous effusion and great congestion.

Being myself of the latter opinion, I had a fine exploring trochar introduced between the eighth and ninth ribs. A little fluid trickled out of the minute canula, immediately followed by the entrance of air into the chest; the body was then inclined to the right side, when about a drachm more escaped, upon which I immediately had the canula removed. The question of adhesion or effusion was thus decided, but how much serum remained and had gravitated towards the diaphragm can only be surmised.

27. *When serous effusion is very considerable, giving rise to unequivocal bronchophony, tubular respiration, and want of resonance, and vocal vibration, physical examination has repeatedly led to a mistaken belief that these signs resulted from pneumonic or other consolidation of the lung.*

A few years ago a little girl was admitted into the hospital, under my care, labouring under great oppression of the chest. Amongst other signs and symptoms, I found great dulness on percussion of the right side, tubular respiration, and so clear and powerful a bronchophony that I did not hesitate to pronounce the case one of pneumonic consolidation. She shortly died, and I found the affected side of the chest full of fluid. More attention to the history and progress of the case would have saved me from this error.

Some time ago a middle aged woman was under my care in the hospital, who had diseased heart and, as I believed, effusion into the right chest. She was seen by another physician, a zealous stethoscopist, who without hesitation and with considerable confidence, declared the case to be one of pneumonic consolidation. I introduced a trocar, and proved beyond all dispute that the pleural cavity contained a large quantity of fluid.

"In reference to the apparent paradox that, in certain conditions of the chest, the vibrations of its parietes, ordinarily produced when the patient speaks, can no longer be felt, whilst, under the same circumstances, loud bronchophony is audible; we must bear in mind that there is no necessary connection between the two results. Sound is transmitted through bodies by a successive motion of its molecules, the whole mass not necessarily undergoing any change of place, whilst the movement appreciable by the hand is the change of place which an entire body undergoes. As a familiar illustration: suppose the end of a deal rod is in contact with the ear, and let a trembling motion be communicated to it, the movement is readily felt by the means of common sensation; but no sound is heard except that of friction, produced by the end of the rod on the ear: here we have motion without sound. Let the rod come to rest, and be firmly applied to the ear, and apply a vibrating tuning-fork to the other end; the sound yielded by it is transmitted by the rod, yet without any appreciable movement of the rod. Not only does the immobility of the rod not prevent the transmission of the acoustic vibrations; but the more firmly the rod is held, the more distinct is the sound. It is true that deep sounds, having also great intensity, do produce a visible vibration of the mass which transmits them; but this arises from the gravity and molecular cohesion of the masses so affected being less, equal to, or only slightly greater than, the acoustic force. If from any cause the body transmitting acoustic vibrations be fixed, there may be no perceptible movement of its mass: such is the condition of the chest, when full of fluid, or affected by pleuritic adhesions; the bronchophony in such cases may be remarkably loud, and yet the hand feel no vibration."*

28. *Even in ordinary acute pleurisy, when the albuminous material thrown out is more abundant; when, in consequence, the lung is held more in contact with the ribs; and when, instead of the whole of the fluid gravitating to the base of the chest, it is more or less confined within the meshes of the solid deposit; it not unfrequently happens that auscultation and percussion fail to determine with certainty whether the physical signs present result from that disease, or from pneumonia advanced to hepatization, or from a combination of the two.*

An instance of this kind has recently been witnessed by some members of the Society in a patient under the care of Dr. Babington.

Were I addressing gentlemen altogether unacquainted with the use of the stethoscope, it might be excusable to point out in detail the means by which we may in general arrive at a correct diagnosis; but, as that is not the case, I shall content myself with offering a caution not to mistake the crepitations so often attendant upon recent pleuritic effusion for the crepitating and mucous râles of broncho-pneumonia; nor to imagine that the physical condition of the chest and physical signs are the same, when the effusion consists

* Dr. Gull.

of pure serum, and when it is made up of the ordinary mixture of serum and solid albumen, usually met with in acute pleurisy occurring in persons of good constitution; the physical signs of the former being, for the reasons stated, much more equivocal and inconclusive than those of the latter.

Although I have just alluded to certain crepitations detected by auscultation, not only in cases of acute, but of ancient pleurisy also —and although I believe that little doubt is now entertained as to their occurrence in both of these morbid states—I do not think it is by any means satisfactorily established whether these crepitations result from the mere movements of the adhesions themselves, or from some mechanical change or impediment in the adjacent lung.

The following communication from my friend Dr. Barlow bears somewhat upon this question:—

"UNION STREET, SOUTHWARK, *June* 26, 1846.

"MY DEAR DOCTOR,

"I much regret that there has not been a complete report of the case of the man who died at No. 3 Lazarus Ward, last Monday, and was inspected on the following day. I trust, however, that the following particulars may serve your purpose.

"This patient was admitted under my care with hæmoptysis, about the end of February last: there were also strong reasons for believing him to be phthisical, although the auscultatory signs were not unequivocal. Soon after his admission he was transferred to the care of Dr. Bird (whilst I had charge of the Clinical Wards) till the end of April, when he again became my patient; and when there were undoubted signs, both topical and general, of advanced tubercular phthisis. Some time before his death, his position, respiration, and the seat of pain, led to the diagnosis of diaphragmatic pleuritis; but there was, moreover, a defined pain and tenderness over a space nearly half-way between the ensiform cartilage and umbilicus, about the size of a crown piece. Here a crepitus could be distinctly felt by the hand; and, upon applying a stethoscope to the part, a crepitation could be plainly heard with each inspiration and expiration. Upon one occasion, a few days before the death of the patient, this crepitation so closely resembled that produced by bronchitis of the smaller tubes, that I made the remark "there is a mucous rattle in the peritoneum." And this resemblance was also noticed by Dr.

Bevan, who was present, and by my reporters, Messrs. King and Rump. Upon inspection, in addition to the disease diagnosed in the chest, there was found a layer of fibrinous effusion on the surface of the parietal peritoneum, corresponding to the situation in which the stethoscopic phenomena had been noticed, and a similar layer on the apposed surface of the liver.

"I am yours, very truly,
"G. H. BARLOW."
Dr. Addison.

29. *When a patient presents himself, with a febrile disorder of any kind, we may, on examination, detect dulness of sound on percussion, tubular respiration, bronchophony, and a râle not distinguishable from that form of mucous or sub-mucous crepitation so commonly observed in the hepatization of acute pneumonia; and yet physical examination shall not enable us to determine whether the chest affection be recent or of ancient date. When a portion of lung has been compressed by pleuritic effusion, and has been prevented from expanding again by surrounding pleuritic adhesions, the physical signs may remain permanently, and be found to resemble precisely those which result from recent pleuro-pneumonia.*

This actually happened in the case of Charlotte C——, already mentioned. There was considerable obscurity about the case altogether; but as she was feverish, and had a frequent hacking cough, and furnished the physical signs just enumerated at the lower and posterior part of the right chest, I scarcely hesitated to conclude that she had pleuro-pneumonia affecting that part. On inspection after death, we found the morbid changes above described of ancient date, and without any trace whatever of recent inflammation in any of the structures involved.

In a case of this kind the stethoscopist is much more likely to err than the man who repudiates the instrument altogether; since the former would be very apt indeed to conclude that recent pleuro-pneumonia existed; whilst the latter, utterly ignorant of the treacherous physical signs, would never entertain the least suspicion of anything of the kind.

30. *Experience leads me to the conclusion that physical examination cannot, in every instance, distinguish the rub and crepitation of*

pleurisy situated at the lowest part of the chest on either side, but especially the right, from similar sounds resulting from recent adhesions between the liver and diaphragm; or between the liver and abdominal parietes: neither can auscultation and percussion at all times distinguish some of the croaking sounds developed in the bronchi from a pleuritic rub; the sounds often closely resembling each other, and both occasionally communicating a manifest vibration to the hand applied to the parietes of the chest.

I met with a case of the abdominal friction sound alluded to above, in a gentleman who died of malignant disease of the liver; and, more recently observed the same thing in a patient whom I attended with Mr. Fidler of Peckham. One of the best illustrations, however, occurred in a patient, Eliza W——, aged 22. This poor creature laboured under a sad complication of disease. She had phthisical disorganisation of both apices of the lungs, an enlarged liver, and the morbid kidneys of Bright's disease. In the progress of her case she was attacked with acute pain and great tenderness over the region of the liver, extending up beneath the margin of the ribs, first of the right, and afterwards of the left side. The crepitation of palpation could scarcely be felt; but a strongly-marked friction sound was heard during the acts of inspiration and expiration, perfectly identical with a pleuritic friction sound. Indeed, it extended so high into the chest on the right side, that it was doubtful whether pleuritis, as well as peritonitis, was not present. Inspection after death proved that there was no recent pleurisy; but "the abdomen contained about a pint of pale straw-coloured serum. The liver weighed six pounds three ounces, extended as high as the third rib, as low as the umbilicus, and across to the left hypochondrium; its upper surface, especially of the right lobe, was coated with a layer of recent lymph; there were also recent adhesions between it and the diaphragm, and the anterior parietes of the abdomen. Intestines slightly reddened, connected together by very soft adhesions, slightly coated with lymph: all these were more apparent on the right side than on the left. Kidneys slightly enlarged, together weighed ten ounces; tunic adherent at some parts, surface irregularly contracted; cortical substance, composed of fatty-looking matter, in which all striated appearance was nearly lost; tubular portion pale; pelvis normal; no fat obtained by heat. Bladder healthy."

31. *As simple pericarditis is rarely attended with pain, and as the other symptoms of that disease are either unsteady or equivocal, the physical signs are chiefly to be relied upon in forming a diagnosis. Nevertheless, when effusion has taken place to a certain amount, the friction sound commonly disappears, and auscultation then fails to recognise the disease.*

The rheumatic ailment of the patient, the more or less distant sound of the heart, the dulness of sound, especially towards the apex of the left lung, and the usually soft and compressible, but slightly jerking, pulse, may perhaps lead to a strong suspicion; but auscultation cannot, in such a case, positively confirm it. Of course, the more considerable and more rapid the effusion, the sooner and more completely does the friction sound disappear; whence the greater difficulty of diagnosis in cachectic and dropsical than in otherwise healthy subjects. I may observe that when, from the quantity of fluid effusion, the ordinary friction sound is no longer heard, by applying my ear to the chest, and retaining it there for some time, I have occasionally been able to detect, once or oftener, a single slight and transient pericardial rub, or rather brush, quite as decisive, in a diagnostic point of view, as the most perfect double pericardial rub.

A curious case of this kind occurred in the person of M——, late porter to the hospital. When requested to see him by Mr. Stocker, he was labouring under severe pleurisy, which was obvious enough; but on applying my ear over the region of the heart, I imagined I heard once such a transient single rub or brush as I have just alluded to. Believing that I had made an important discovery, I carefully repeated my exploration; but although I continued to listen a considerable time, I could never hear it again. He presently died; and on examination we found universal and intense pericarditis, with unusually copious sero-albuminous effusion.

32. *Enormous accumulations of fluid in the pericardium cannot by physical signs be distinguished at all times from effusion into the cavity of the pleura.*

33. *When the pericardial friction sound is single, auscultation may fail to distinguish it from a valvular murmur; and especially so when the single friction sound is heard most distinctly over the situation of the valves.*

34. *So far as auscultation is concerned, the double friction sound of pericarditis, if heard toward the base of the heart, might be mistaken for the see-saw murmur of imperfect aortic valves, or* vice versâ.

The character of the pulse alone would be almost sufficient to decide the question.

35. *Auscultation does not at all times enable us to distinguish a friction sound produced within from a friction sound produced without, the pericardium; i. e. friction between the pericardial surfaces from friction between the loose pericardium and lung or parietes of the chest.*

I have had more illustrations than one of this difficulty; and the following case, admitted into our Clinical Ward, I am disposed to think furnishes another.

"Samuel F——, aged 21, was admitted on Sunday night, the 7th of June, in a state of partial collapse. He had been seized about two hours after his meal with excruciating pain in the lower part of the abdomen: there were frequent, but ineffectual, efforts to vomit. The pain extended so as to involve all parts of the abdomen, which was exquisitely tender on pressure. He gradually sank into a semi-comatose state; the extremities became cold; respiration was hurried, and effected with great apparent pain; the pupils were contracted; consciousness of surrounding objects was imperfect.

"*Diagnosis*—Perforation of the stomach.

"The chest being examined, gave the following stethoscopic signs:—Over the aortic valves was heard a soft brushing to and fro sound, accompanying each systole and diastole of the heart. The sounds of respiration anteriorly were normal. Behind, at the base of the right lung, a harsh grating rub met the ear, synchronous with inspiration and expiration: elsewhere, breathing was normal.

"The patient sank rapidly, and died the following day at twelve o'clock.

"*Sectio-cadaveris.*—A perforating ulcer in the lesser curvature of the stomach. Universal peritonitis.

"Lungs and heart perfectly healthy. Between the right pleura and pericardium there existed a band of old adhesions, which had in all probability given rise to the anomalous *bruit* heard with the heart's action. Both lungs overlapped the heart.

"The peritoneal coverings of the liver and diaphragm were coated with recently-effused albuminous deposit. The ascent and descent of the diaphragm during the respiratory act, bringing these two surfaces into contact, had caused the rubbing to and fro sound heard at the base of the right lung."

36. *A sound closely resembling a valvular murmur appears not unfrequently to be produced by the stroke of the heart against a portion of lung, interposed between it and the parietes of the chest. Under such circumstances, auscultation may lead, and I believe often has led, to the erroneous conclusion, that the heart is diseased, when it is perfectly normal in every respect.*

This sound is most frequently heard at some point in the direction of the edge of the left lung, where it overlaps the heart, to the left of the sternum, from about the second or third to about the fifth rib, and especially somewhere between the second rib and the neighbourhood of the left nipple. Its tone somewhat resembles that of a *bruit de rape*; but, at the same time, it communicates a sense of dryness and crumpling, different from the rigid squeezing or grating observed in the ordinary *bruit de rape*. It is also more variable, both in its development and its extent. We find it different at different moments, and during the different movements of the chest; and it may occasionally be made to disappear altogether, by a deep and forcible inspiration, so long as that inspiration is maintained by the individual. On the other hand, its extent or prolongation varies in different cases; or even at different times in the same case; apparently according to the extent or size of the portion of lung which happens to be struck by the heart at each systole of the organ. In a few instances I have found the sound, to a certain extent, double; the second, or that attending the diastole of the heart, being in general, perhaps, more limited and indistinct than the first. I believe this sound, which I have long observed, and now attempt to describe, to be that recently pointed out by my colleague, Dr. Barlow. It may possibly be that also noticed also by Dr. Latham as frequently present in phthisis.

I have met with it occasionally under the clavicles, resulting probably from the impulse of the aorta or large arteries; although at first, I was led to entertain serious apprehensions of the existence of aneurism.

PHYSICAL DIAGNOSIS IN DISEASES OF THE CHEST. 93

Instances of extrinsic cardiac murmurs are, in truth, exceedingly common, and ought to be carefully borne in mind in exploring the chest. The following case is probably one of this kind.

"JOSEPH C——, aged 24, was admitted into our Clinical Ward on the 8th of July. He complained of dyspnœa, and a painful feeling of weight at the bases of both lungs when a deep inspiration was taken. The history was that of recent supervening upon old pleurisy. The stethoscopic signs indicated phthisical disease at the apex of the right lung, pleuritic adhesions below the left mamma, and consolidated lung with pleurisy and dilated bronchial tubes at the left base posteriorly. The heart's position in the chest, its impulse, and rhythm, were perfectly normal; no *bruit* accompanied either sound.

"But little change, save that of a gradual extension of the pleurisy, took place in the general or physical signs, until the 24th. On this day, the ear being placed over the aortic valves, a rubbing sound was heard accompanying the heart's action. Inasmuch as there was no symptom whatever of any affection of the pericardium, the nature and cause of the *bruit* underwent a rigorous investigation. It was found not to be persistent, but to vary as the act of respiration varied. The patient was ordered to take a full inspiration and hold his breath. The heart's sounds were then heard, somewhat feeble and distant, but perfectly normal. The lungs were then allowed to collapse partially, and the breath was again held. A distinct double *bruit* became audible, accompanying the systole and diastole of the heart; a loud, harsh *bruit de rape* with the systole; a soft brushing sound with the diastole. The act of expiration being fully effected and the breath once more held, the heart was heard to beat loudly and distinctly, and nothing abnormal could be detected in its sounds. —The patient is still in the hospital, and the phenomenon persists unchanged."

For the particulars of the next case, I am indebted to my friend Dr. Gull.

"W. H——, aged 27, admitted, under Dr. Barlow, April 22, 1846: engineer by employment; single; very muscular; habits irregular. The first symptoms of his present disorder came on about three days previous to his admission; his legs at that time beginning to swell. For a few days after his admission, he went on as usual with cases of renal anasarca; the urine being loaded with albumen, but not tinged with blood. From some imprudent exposure he became feverish; great dyspnœa supervened; his urine was scanty, and seemed little else than pure blood.

"There were signs of universal bronchitis, with imperfect respiration, and indistinct tubular breathing at the base of the right lung. The next day, his dyspnœa continued very urgent: great pain across the præcordia, extending into the left side: impulse of heart indistinct, with excessive dulness; sounds indistinct. On passing the stethoscope up to the space between the second and third costal cartilages, a well marked to and fro sound was audible, continuing during the suspension of respiration. He was too ill to examine the chest generally.

" Added to the above stethoscopic indications, regarded as indicative of fluid into the pericardium, was his waking from sleep in a sudden fright, his constant erect position in bed, the great feeling of oppression about the heart (to use his own words) 'as if a cupping-glass were pressing over the spot.'

" The reasoning on the case was something in this manner :—He has renal disease, of an acute character; nearly entire suppression : we have the usual renal catarrh of the bronchial tubes, pleuritic effusion of right side, acute pleurisy of left side, and pericarditis. In these renal cases we have a considerable quantity of fluid with the plastic matter: from his erect position, the more fluid parts would gravitate, leaving the surfaces of the pericardium in apposition above, where there is audible a distinct to and fro sound : now this may be pleuritic, but it continues unaltered when the patient holds his breath. Thus again the indistinctness of the sounds and imperfect impulse would coincide with fluid in the pericardium. There seemed but one sign opposed to such a view, and that was the large volume and firm character of the pulse, which peculiarity seemed explicable on the well-known concomitance between hypertrophy of the left ventricle and renal disease.

" *Sectio-cadaveris.*—General œdema of the body; the results of acute inflammation in the left side, the pleura containing a considerable quantity of serum, turbid from flakes of fibrin. The pleura was covered with a layer of plastic lymph; and between the second and third ribs the fibrin was so situated on the pleura pulmonalis, as to be influenced by the movements of the heart, and thus produced the to and fro sound, which could be artificially produced at the post-mortem examination by traction of the pericardium. The pulmonary tissue of the left lung was in the first stage of pneumonia (œdema). On the right side, the upper lobes of the lung were emphysematous; the lower lobes were devoid of air, from compression of the clear serous fluid in the pleural cavity; no signs of recent inflammation. The bronchial tubes, on both sides, contained a large quantity of viscid mucus.

"Heart.—Right side very much distended and pushing the left side backwards : (did not this render the impulse so indistinct, and cause the increased dulness?) : left side hypertrophic : weight of heart, fourteen ounces and a half. Several ounces of clear serum into pericardium.

"Liver enlarged and congested; weight, six pounds one ounce.

"Spleen.—Weight, fifteen ounces; granules indistinct.

"Kidneys very large; weight, twenty-five ounces and a half; dark chocolate; tissue soft."

A very curious modification of extrinsic cardiac murmur occurred in Miriam Ward a few years ago. A young female, with diseased lungs, furnished, on auscultation, a double murmur of the character described. The sound was very strikingly developed, both in the systole and diastole of the heart; but mixed up with a remarkable moist sucking sound, somewhat resembling what would be produced by the application and up-and-down movement of the leathern plaything called a "sucker." The diagnosis formed was vomica, with adhesion to the pericardium; and this was found to be correct on inspection after death.

The only doubt remaining in my own mind in this matter is, whether the extrinsic cardiac murmurs, which I have been in the habit of calling *pulmonic,* may not in every instance be associated with some slight or circumscribed adhesion, or albuminous deposit.

37. *Auscultation fails to distinguish an aortic murmur depending upon organic change from that which results from other causes.*

The murmur from diseased valves cannot always be distinguished from the murmur so often attendant on chlorosis, diseased spleen, or other form of anæmia. The general history, coupled with the character of the murmur, will very commonly decide the question.

38. *Auscultation alone cannot determine whether what has been called a mitral murmur result from organic or functional change.*

Auscultation cannot distinguish the mitral murmur of endo-carditic disease from the temporary mitral murmur occasionally met with in chorea.

39. *No physical examination will enable us, in many instances, correctly to analyse the morbid condition of a heart, whilst that organ continues to be greatly overcharged with blood.*

Hence the prudent rule, always to wait a certain time before giving a positive opinion. This precaution applies more especially to cases in which severe bronchitis complicates the cardiac affection.

40. *In certain diseases of the heart, especially when the organ is enlarged, it is difficult, or impossible, accurately to localize the murmurs, however distinct and obvious these murmurs may be.*

The two following cases have recently come under our observation; and although, perhaps, not very strongly marked, are, nevertheless, instructive.

" James T——, aged 56, a tanner by trade, residing in Bermondsey. A year and a half ago, after exposure to bad weather, he became affected with cough, attended with expectoration and hoarseness of voice. These symptoms continued, sometimes aggravated and sometimes remitting in severity, till thirteen weeks ago, when he became much worse, suffering from occasional attacks of dyspnœa and œdema of the extremities.

" On admission, the positive physical signs of thoracic disease were as follows :—Dulness, extending from the left nipple to the margin of the ribs; bounded by the sternum to the right, and posterior line of the left lateral region, to the left. Absence of vesicular murmur over the same space. Large mucous crepitation heard irregularly over various parts of the lungs. *Action of heart very irregular; frequently intermittent;* contraction feeble. Every now and then a stronger beat is made; and at such times a loud *bruit* is heard with the first sound. This *bruit*, perhaps loudest over the aortic valves, may be traced almost as far down as the margin of the ribs, and along the great vessels up to the clavicles. It is difficult to say in which direction it is heard more distinctly. Pulse 100.

"The condition of the heart, on post-mortem examination, was the following:—Great hypertrophy of both sides; weight, twenty-four ounces: right and left auriculo-ventricular orifices large: pulmonary artery thickened, having very much the appearance of the aorta; its valves healthy: mitral cords stretched; curtain thickened somewhat; aortic valves rigid from a large deposit of bony matter, extended and leaving a mere slit for the passage of the blood. The pulmonic signs discovered by auscultation, which had led to a suspicion of effusion into the left chest, were all due to the great hypertrophy of the heart."

"JAMES A——, aged 24, ill four years: a sailor, unmarried, living at Woolwich. He had never had any serious illness until four years before admission, when he fell into the water, and had an attack of rheumatism, which lasted two months. After this, on making any exertion, he felt palpitation, dyspnœa, sensation of weight in the chest, with pain and tenderness at the epigastrium.

"When admitted under Dr. Addison, February 4, 1846, his aspect was tolerably healthy; he had a slight cough, with white, scanty sputa. *With the exception of an occasional intermission, the pulse was regular.* Impulse of heart and præcordial dulness increased; mucous râles in various parts of the chest; *bruit* on first sound of heart below mamma, second sound scarcely audible : there was severe pain in the loins, and pain and tenderness at the epigastrium. It was at first supposed that the symptoms were produced by simple hypertrophy and dilatation; but shortly afterwards it was discovered that the *bruit* was loudest at the epigastrium, where a tolerably well-defined pulsating tumour was felt. The *bruit* could not be heard posteriorly, and was constant in all positions of the patient. The case was now supposed, by Dr. Addison, to be either aneurism of the abdominal aorta; or adhesion between the cardiac and diaphragmatic pericardia. The symptoms which ensued, seemed to confirm the former opinion. The cough became loud, frequent, and violent; the patient declaring that he was trying to cough up something which he felt in his throat. The food appeared to stop at the epigastrium before it passed into the stomach; and on two or three occasions it regurgitated soon after it had passed down the gullet. The voice became thicker; he felt choked when he lay on his back;

pain and oppression at the epigastrium increased. During this period the *bruit* and pulsation remained unaltered. On the 4th of July he began to cough up blood, which gradually increased in quantity: and on the 18th of July he died.

"*Sectio-cadaveris.*—Apoplexy at the base and apex of the right lung posteriorly, and, to a less extent, at the posterior part of the left lung: pericardium connected to the heart by old adhesions; right auricle dilated; right ventricle hypertrophied and dilated; tricuspid valve opaque, thick, pliable, with small granulations in its auricular surface; pulmonary crescents healthy; left auricle large; some small bony projections on auricular surface of mitral; orifice of mitral greatly contracted; left ventricle dilated, its walls of usual thickness; mitral valve rigid and opaque, its cords contracted, thick, and firm. The thick, hard mitral formed a projection into the ventricle, which pressed against that smooth track of lining membrane which leads up to the aorta, the membrane at the point of contact being thickened and hard. Small granular excrescences on the ventricular surface of the aortic valves: slight atheromatous deposit in the descending aorta, and also in the pulmonary arteries, which were thickened and enlarged. The circumference of the aorta, just above the valves, was 2·63 inches. No aneurism, nor any abdominal tumour, pressing on the aorta."

41. *Auscultation cannot distinguish the murmur of an aneurismal or otherwise diseased artery, from the murmur occasioned by some source of pressure upon the same vessel.*

42. *Physical examination does not enable us to distinguish congenital malformation from disease of the heart or large vessels.*

There are, however, many exceptions to this proposition.

To have treated fully the many important questions comprised in this communication would have rendered it totally incompatible with the means and object of the Society. This must plead my excuse for having preferred the greatest possible brevity to more elaborate detail. My object has been to acknowledge, and point out, the many difficulties and fallacies which I have myself had to encounter in the practice of physical diagnosis, without presuming, for an instant, to suppose that they would have proved equally embarrassing to other stethoscopists.

OBSERVATIONS

ON

FATTY DEGENERATION OF THE LIVER.

This morbid condition of the liver has hitherto attracted but little attention in this country; the extremely imperfect information which we at present possess concerning it having been obtained from the French physicians; whilst even amongst them, if we accept its imputed connection with phthisical disease, it must be confessed that no advance whatever has been made towards a knowledge of either its causes or its consequences. This want of information is probably, in a great measure, attributable to our inability to recognise the disease during the life of the patient; for as we then entertain no suspicion of its existence, we fail to make the minute and connected observation of facts so necessary to the elucidation of every internal affection.

We find the utter barrenness of diagnosis upon this matter well illustrated in the candid avowal of the accomplished and experienced M. Louis, who, in his admirable work on Phthisis, observes: " Nous manquons de signes capables de la faire reconnaître à une époque quelconque de sa durée. En vain nous avons été au-devant des symptomes qui pourraient lui appartenir, nous n'en avons recueilli aucun."

To supply this defect ought, then, to be the endeavour, as it unquestionably is the duty, of every one enjoying the advantages of a large hospital, and it is in this spirit that I submit to the profession the scanty materials of this communication.

When the liver has undergone the extraordinary change in question, it exhibits, when the abdomen is laid open, a pretty uniform and highly characteristic appearance. It is observed to be of a cream or pale-yellow colour, figured irregularly with brownish or deep-orange spots. It is usually, though not always, more or less enlarged, and sometimes very considerably so.

When cut into, its interior is found to present an appearance somewhat corresponding to that of the exterior, excepting that the brown and pale-yellow tissues are much more uniformly distributed throughout the organ than they are upon its surface. It is sometimes softer, and more readily crushed between the fingers than is the healthy liver; sometimes, however, it is firmer than natural, and occasionally even a scirrhous or almost horny hardness. It will often very perceptibly grease the scalpel, or impart an unctuous feel to the fingers; and on being exposed to the flame of a candle, will yield a considerable quantity of fat. Although its absolute weight, when enlarged, may exceed that of the healthy liver, I am disposed to think that its specific gravity will, in perhaps the majority of instances, be found to be less: thus a liver apparently healthy was of the specific gravity 1·0625; whilst the specific gravity of two fatty livers, examined at the same time, was only 1·0275. I have, moreover, recently met with a case in which the liver was not only much enlarged, but exceedingly dense and hard, and consequently, as I suppose, of great specific gravity.

The following analysis of the liver, bile, and urine of a patient whose liver presented, after death, the morbid condition under consideration, was furnished me by my friend, Mr. Golding Bird.

"The bile differed from the healthy form of the secretion in its colour, which was of a dirty brown, scarcely partaking anything of a green tint; it contained in diffusion, numerous black grains, which were insoluble in water, alcohol, and alkalies; burnt without flame; detonated when thrown on fused nitre, causing the formation of carbonate of potass: hence these granules consisted of carbon nearly pure, which substance has been, I find, detected floating in the bile of a maniacal patient by Prof. Lavini of Turin ('Atti dell' Academia de Turino,' 1834). Numerous globules of oil appeared after repose, on the surface of the bile; but their quantity could not be readily determined: in other respects the secretion presented the same chemical characters as ordinary specimens of bile. Its composition was as follows—

Biliary matter, with an appreciable quantity of very soft fat	6·0
Earthy salts (phosphates of lime and magnesia)	0·2
Osmazome, lactate of soda and common salt	1·0
Water	92·8
	100·

The most remarkable deviation from the healthy state apparent in this specimen of bile was the peculiar odour it evolved after the addition of an acid, even of acetic. This odour was the most disgusting and least tolerable I ever met with evolved from any kind of animal product, it was quite *sui generis*, and very permanent.

The liver was next examined for the purpose of ascertaining the quantity of fat present. One thousand grains were cut into very small pieces, and boiled in repeated portions of alcohol, the solutions being filtered, whilst nearly boiling hot, through muslin. By this process a solution of fat and osmazome, with saline matter, was obtained: the fluid was evaporated over a vapour bath, to dryness, and the residue digested in boiling water, to remove the salt and osmazome. The insoluble portion consisted of a soft brownish fat, very fusible, and possessing a peculiar and unpleasant odour: the quantity yielded by the one thousand grains was 47·6 grains, equal to 333·2 grains in each pound of liver.

The urine of the same patient was examined on June 1st: it was remarkable for its extremely low specific gravity, being but 1·007: it was quite neutral; contained in suspension dense mucous flocculi, and yielded a precipitate by protracted ebullition, which was insoluble in nitric acid; it however differed from albumen, in not being *precipitated* by nitric acid. It contained but very minute traces of the earthy phosphates. When evaporated to dryness, it left a residue, soluble partly in alcohol, and remarkable for its peculiar bitter taste, which I had never previously met with in any specimen of urine. That this bitterness is not owing to the presence of bile, was evident from its being scarcely coloured, and not yielding any change of colour, on the addition of nitric acid."

Having, in the course of my experience, been often struck with a remarkable appearance of the face in certain patients—an appearance dependent not so much on the expression of the countenance as the texture and aspect of the integuments—and having observed the exact resemblance of the appearance in each case, I endeavoured to connect it with some corresponding uniformity in the accompanying disease; and at length arrived at the conclusion, that, *when strongly marked, it is indicative, if not pathognomonic, of fatty degeneration of the liver.* It is purely integumental, as it is not confined to the face, but may pervade the whole surface of the body; although I am not disposed to think that it is earliest observable, as well as most conspicuous, in the integuments of the face and backs of the hands. To the eye the skin presents a bloodless, almost semi-transparent and waxy appearance; when this is associated with mere pallor, it is not very unlike fine polished ivory; but when combined with a more sallow tinge, as is now and then the case, it more resembles a common wax model. To the touch, the general integuments, for the most part, feel smooth, loose, and often flabby; whilst in some well-marked cases, all its natural asperities would appear to be obliterated, and it becomes so exquisitely smooth and soft as to convey a sensation resembling that experienced on handling a piece of the softest satin. Whether this condition of the integuments precede or follow that of the liver, and whether the two are necessarily associated in every instance, I am by no means prepared to offer a decided opinion, but having on several occasions confidently and correctly predicted finding the fatty liver, from having observed the particular condition of the integuments just described, I have ventured to make the crude fact known to the profession, in the hope, that, attention being once directed to such a starting point, something more interesting, and of greater practical utility, may result from its further investigation.

M. Louis, in the elaborate work to which I have already alluded, dwells upon the frequent occurrence of this degeneration of the liver, in connection with tubercular disease of the lungs. He even estimates that it occurs in one third of the whole number of phthisical patients in France; as in 120 cases, he met with 44 instances of the complication; and he even goes so far as to regard it as a mere part or consequence of such tubercular disease. He appears to have been led to this conclusion, by observing, that in almost every case of fatty liver there existed more or less indication of tubercular disease in the lungs; and that in the development of the fatty degeneration of the liver it is not materially influenced either by the age, sex, or constitution of the patient; except, inasmuch as these circumstances affect the phthisical disease, upon which he considers it to depend.

But although this condition of the liver is most frequently found associated with tubercles in the lungs, this is far from being uniformly the case, in this country at least: and it is not a little remarkable, that within the short period of the last two months I have myself met with no less than three instances, in which not the slightest vestige of tubercular disease was discovered, either in the lungs or in any other part of the body: and it is further worthy of notice that in two of them (Elizabeth Castline, aged 32, and Samuel Minx, aged 57) no suspicion had, to my knowledge, been entertained of disease existing, either in the lungs or liver; both individuals having remained plump and fat up to the period when death took place. Elizabeth Castline died of organic disease of the brain, and the following is the account of the appearances found on dissection, furnished by Mr. King, who examined the body :—

"There was a nævus in the pia mater, near the vertex; and the ventricles of the brain were distended, with two ounces of pretty clear fluid.

"The contents of the chest were tolerably healthy, and the blood in the heart was coagulated and fibrinous.

"The liver was of a large size, rather pale, coarse, and close, and yielded much oil when baked in paper. Bile pretty copious, and dark yellow, nearly to blackness. The spleen was of good size, and coarsely granular. Uterine system in a healthy condition."

S— M— died of pericarditis, succeeding to an attempt to commit suicide by cutting his throat.

His body when examined was much distended by the gaseous products of decomposition : the lungs were watery, and the pericardium lined with honeycomb fibrin : the spleen unhealthy ; and the liver fatty, and presenting the usual characters of that state.

Whether any symptoms or indications attended this condition of the liver in these two individuals, or what they were, I had no opportunity of ascertaining; but what has been stated renders it sufficiently apparent that fatty liver is not necessarily associated with tubercular disease of the lungs ; and that such a state of liver is not, for a time at least, incompatible with a certain degree of bodily vigour.

The following case, which was furnished me by Mr. Barnes of Chelsea, presents an instance of an asthenic or atrophic state connected with this change of the liver; which I am inclined to believe is of by no means unfrequent occurrence, and tends to show not only that fatty degeneration of the liver may exist, unconnected with the tubercular disease of the lungs, but that its existence may occasionally be pretty confidently diagnosticated during the life of the patient.

"In January, 1835, I first saw Mrs. T—, who was then suffering from a very troublesome cough, which she said was usual at that period of the year; her countenance was anxious; great emaciation, and a total loss of appetite, had taken place; her pulse was very feeble, and under 70: tongue clean: bowels generally regular; and evacuations healthy. There was no expectoration, and no perspiration. From her general appearance, I thought she was phthisical auscultation, however, afforded no signs of disorganization; nor was her pulse such as would lead one to infer that tubercles existed. This cough continued more or less till May, when it left her. During my visits, from January to May, she occasionally suffered from sickness, which was always relieved by prussic acid; she had rarely occasion for purgative medicine. From this period till March, 1835, I heard nothing of her, and concluded she must have been under the care of some other person. Such was not the fact; she had occasionally suffered from cough, but not to any extent: her emaciation continued, and her appetite had not increased. To all appearances, no change had taken place, her spirits were greatly depressed, having lost a sister a few weeks before.

"At this time, auscultation elicited nothing satisfactory.

"The medicines which afforded relief before were now taken without any good results; and I requested permission to call in Dr. Addison.

"At that time she was able to move about, complained of great weakness, but no pain; indeed, she said she had never felt pain. Her hands at this time began to swell, towards evening; and there was also some puffing of the face. Her urine had always been plentiful, and did not appear changed: it contained no albumen.

"Dr. Addison examined the chest with great care, but could not make up his mind if disease existed there; as the stethoscopic signs led to the belief that there was no change of structure going on, whilst the general symptoms seemed to favour a contrary opinion.

"Three weeks after this, œdematous swelling of the feet commenced: and finding that it had disappeared in the morning, she could not be prevailed upon to get up, for fear it should return. It may be observed that she never experienced any difficulty in breathing; had no occasion to have the head or chest raised; and could lie on either side. Dr. Addison saw her again in a short time; he could perceive no change, excepting her increasing debility, which was occasioned by her refusing all food; her only support being a little weak wine or brandy and water. She slept tolerably well. The treatment during this consisted of small doses of blue pill, com-

FATTY DEGENERATION OF THE LIVER. 105

bined with extract of henbane and a grain of quinine night and morning, and a drop or two of prussic acid with a few grains of magnesia in camphor mixture : she went on thus till the end of May, when she died.

"Having obtained permission to examine the body, I requested the attendance of Dr. Addison, who had always said that this patient had a fat liver, but had been unwilling to venture a decided opinion respecting disease in the chest, the impression, however, on his mind had been that little, if any, existed.

"Having opened the chest, we found extensive firm, old adhesions of the pleura, but we could not discover a vestige of tubercular disease in the lungs, even upon the most careful examination; they were, indeed, to all appearance, quite healthy; excepting that, like other parts of the body, they were easily broken down.

"The heart was sound.

"When the abdomen was laid open, the liver was seen somewhat enlarged, presenting a straw colour in appearance, and rather firm. We cut a thin slice from it; and having exposed to the flame of a candle, it did not curl as is usual with healthy liver; but it gave out a considerable quantity of fat, which was allowed to drop on paper; the stomach was, to all appearance, healthy, excepting the mucous covering being rather hard in the follicles, and somewhat of a horny aspect. The kidneys, uterus, and bladder were examined and had a healthy appearance."

In this instance I was led to form the diagnosis regarding the state of the liver entirely from the aspect and feel of the integuments especially on the face and backs of the hands, for during the life of the patient she had exhibited no decided symptoms of disease of that viscus; and, indeed, the character and situation of the anasarca under which she had laboured led me to suspect some morbid change in the kidneys, but on examining the urine no trace of albumen could be detected. I think it not improbable, therefore, that this degeneration of the liver may, like mottled kidney, occasionally prove a cause of anasarca; a supposition somewhat strengthened by finding, on visiting her shortly before death, a great degree of anasarcous infiltration of the back, loins, and sides, exactly such as I have often found in renal diseases, although with much less swelling of the legs and thighs than usually accompanies the latter. In one case it scarcely existed at all, or at least in a very slight degree. Could this anasarca be a mere consequence of inanition; the appetite being as it were completely obliterated? Certain it is that there was not only remarkable flabbiness of the soft solids during life, but, as I have observed in more cases than one, an extraordinary lacerability and want of tenacity of the tissues generally, and of the lungs in particular, after death. As bearing upon this question, however, I may add, that I have recently seen a case in which anasarca had occurred, and again disappeared, long prior to the death of the patient. This person was a female of the name of Gowan, aged 30, under the care of Dr. Bright, in Lydia's Ward. On passing through the ward some time before her death, my attention was attracted by the peculiar aspect of the integuments already described; and on examination of the chest, I found that she was sinking from phthisical disorganization of the lungs. Convinced that she had a fat liver, notwithstanding her stools presented a natural appearance, I interrogated her very narrowly, and ascertained that several months before, she had had anasarca, which disappeared under medical treatment. Upon dissection the lungs were found studded with miliary tubercles, and there was a vomical cavity. The liver was fatty.

With respect to the causes of this fatty degeneration of the liver, very little, or absolutely nothing is known. In most of the cases which I have met with, there has been either positive or strong presumptive evidence that the individuals had indulged in spirit-drinking, and indeed the most exquisite case I ever saw in a young subject, occurred in a female who had for some time subsisted almost exclusively on ardent spirits. On the other hand the extreme frequency of the degeneration in France, where the people are but little given to such indulgences, throws considerable doubt upon such an origin of the complaint, whilst its great frequency there in conjunction with phthisis, would almost lead to a belief that it has some connection with a scrofulous tendency, a belief which, I confess, I am strongly disposed to entertain.

Connected with scrofula or not, it remains to be ascertained whether it be in reality an original disease of the liver; or whether it may not, in every instance, be merely secondary to some other or remote disorder, is not yet determined. Future experience, also, must discover what are its immediate, and what are its remote consequences, whether feebleness of the powers generally, and of the digestive organs in particular, be sooner or later induced by it, in every instance where the patients escape other mortal diseases; whether, by impairing the powers of the constitution, it may not, in some instances, prove rather a cause than a consequence of phthisis; whether anasarca or any other form of dropsy be peculiar to certain stages or degrees of it; and, lastly, what influence it exerts upon the brain and nervous system generally, and consequently upon the feelings and disposition of the individual.

DISORDERS OF FEMALES CONNECTED WITH UTERINE IRRITATION.

PREFACE.

HAVING been honoured with a share of the Clinical duties in the Medical School of Guy's Hospital, I availed myself of a few Cases admitted into the Female ward, to illustrate, to the best of my ability, a very prevalent and very important class of diseases which had long engaged my attention, a class of diseases most seriously affecting the health and comforts of Female life, and consequently fraught with the deepest interest to the Profession. It was after the delivery of a general Lecture on the subject that I was recommended by a few friends, upon whose candour I thought I could rely, to publish my opinions. To this, I can truly affirm, I felt considerable reluctance. I was too much occupied in the performance of laborious public duties to compose such a work as might be deemed worthy the acceptance of the Profession, whilst I had little inclination to intrude upon their notice so slender a production as a Clinical Lecture. It is of very little consequence how or why I overcame such scruples:

suffice it to say, that the opinions advanced were adopted with caution, and have stood the test of several years of the veriest professional drudgery. If the merit of these opinions be small, it is the more commensurate with my pretensions; if I fail to multiply the means of relief, let me at least indulge the hope that I may have diminished the sources of error. To the experienced practitioner the contents of this little pamphlet may appear as trifling as its form and style are unattractive and homely; but, should it contribute in any degree to preserve the Junior Members of the Profession from those errors into which I myself have fallen, I shall not regret having permitted its publication.

T. A.

24, NEW STREET, SPRING GARDENS,
    March 13, 1830.

ON THE

DISORDERS OF FEMALES CONNECTED WITH UTERINE IRRITATION.

THE object of this lecture is to make you acquainted with the morbid effects produced upon the general constitution, and upon particular parts of the body, by continued *Uterine Irritation*. In making the attempt, some apology is probably due to my excellent friend and colleague, Dr. Blundell, for thus encroaching upon his particular province. But, gentlemen, if I am correct in attaching an unusual degree of importance to the subject, and can succeed in imparting any really useful and practical knowledge to you, I am too sensible of the zeal, intelligence, and liberality of my friend, to entertain the least apprehension of giving him offence.

There are few circumstances more powerfully calculated to excite serious reflection in the breast of one acquainted with the history of our profession, than a knowledge of the various opinions that have at various times prevailed respecting the nature of disease in general; opinions which, like disease itself, have been observed to have their rise, height, and decline; opinions which have for the most part held sway, or been revived, rather in deference to the authority of a great name, than to the results of legitimate and dispassionate investigation. When the Egyptians attributed disease to the influence of demons of the air, they, perhaps, did a very foolish thing; but before we consign them to ridicule and contempt, we ought, in justice, to award to them the merit of having been just as intelligible, and quite as reasonable, as many of the reputed lights of after-ages. Hippocrates had his *nature*, his *humours*, and his *concoction*; *Asclepiades* had his *pores* and *corpuscles*; *Themison*, his *stricture* and *relaxation*; the *Galenites* added their own absurdities to those of Hippocrates, and kept possession of the medical schools for 1500 years, whilst in more modern times, fancy has equally run riot, and furnished ample material for the strife of opinions.

It may indeed be alleged in reply, that such evils, such imperfections, were not absolutely inseparable from physic, but resulted rather from the deplorable ignorance of those who professed it. Now, gentlemen, this may be true, yet with all the advantages we at present possess, with all the improvements of our own times, we still have to deplore the very imperfect state of medical science. The functions of the human body are so numerous and complicated, they are so intimately connected with, and dependent upon each other, and their derangements are so diversified according to age, constitution, and other circumstances that, amid the obscurity attendant upon all vital phenomena, it would be little short of a miracle were all observers precisely to agree in the various relations of cause and effect. Hence the diversity of opinion prevailing even at the present day, in diseases, too, of repeated and daily occurrence, and amongst men equally meritorious, and equally deserving our respect.

The distinctions of *general* and *local* diseases, at one time undisputed and apparently well, or at least boldly defined, have in our days, undergone such an ordeal as to leave us in doubt of their very existence; whilst we have seen arising from the ashes of former opinions another "loop to hang a doubt upon;" we have seen one party contending for the *constitutional* origin of *local* diseases, another party, with equal plausibility, contending for the *local* origin of disorders formerly believed to be *general*; whilst a very numerous class of important diseases constitute at the present day a sort of neutral ground, concerning which there appears to exist a truce amongst all parties; a truce, however, only to be observed until the decision of present contests shall afford time and opportunity to break it. Thus, *fevers* were from time immemorial regarded as *general* diseases, but are now traced, or are attempted to be traced, to originally *local* disorder, the champions of the cause being *Clutterbuck*, in this country, and *Broussais* in France. On the other hand, various diseases, formerly regarded and treated as purely *local*, have been traced to a *constitutional* origin; the great champion of the cause being our distinguished countryman, Mr. Abernethy; whilst the *phlegmasiæ*, as they are called, form the neutral ground of which I have spoken—a neutral ground, which, like Poland of old, remains in reserve, as a prize to the victors; for, although tacit, but reluctant assent refers them to the class of *local* disorders, doubts and disputes have ever and anon crept in to disturb the creed. Individuals have observed, or have imagined they have observed, that in many of these phlegmasiæ, in many of these reputed *local* inflammations, the *general* or febrile symptoms often anticipate the *local* inflammation, and consequently cannot be truly said to be secondary or symptomatic of that local inflammation; neither are the *general* symptoms at all times in proportion to the degree of the *local* inflammation, which ought to be the case, were the general symptoms purely symptomatic, as may be repeatedly witnessed both in catarrh and in rheumatism.

Now, all this abundantly attests the many difficulties which oppose themselves to the attainment of truth; it teaches us humility in the pursuit; it furnishes an apology for the many errors we all of us commit in the exercise of a profession, founded upon general and ever varying principles—principles the result rather of multiplied observation and experience, than deducible from any distinct knowledge of the vital endowments of the human frame. But, gentlemen, this diversity of opinion in regard to the constitutional or local origin of disease, teaches us, at the same time, a most impressive lesson as to the intricate and intimate connexions existing between the different systems and organs of the living body;—it proclaims, in language which cannot be misunderstood, the danger, the folly of reasoning upon the functions of the living body, as if it were a mere piece of mechanism; and confirms the important truth, that, whether we call a disease general or local in its origin, the sympathetic or secondary effects which arise during its progress, will vary and require correction widely different from what we observe in any derangement, however complicated, of wheels and levers. If a mere piece of mechanism go wrong, from some defect in one of its parts, although it may derange the movements of the whole, so long as the partial defect is allowed to remain, yet it will generally be found, that it is only necessary to correct that original defect to restore the whole to regular order; not so with the living body, for the secondary effects, as well as the primary disorder, must be combated, in order to restore the entire machine to a healthy condition : nay, so true is this, especially in what we call *idiopathic fever*, that the secondary or sympathetic affections often demand our special and almost exclusive attention, whilst they constitute the chief sources of danger and alarm. Again, let us reverse the position, and let us suppose a person to receive a severe *local* injury from external violence, we shall presently have the whole system participating in the disturbance, and perhaps to such an extent, as to constitute by far the most important and most prominent feature of the case.

Now, in the instances quoted, the sympathetic or secondary disorders are so manifest, and are so uniformly present, that so far from questioning the connection, we look for them as a matter of course, and, in short, are apt to regard the combination as merely constituting one entire disease. But, suppose the general disorder of the frame to be tardy and insidious in its approach; suppose it, from the nature of the offending cause, to steal almost insensibly upon the patient, its presence not being indicated by any signs or symptoms, very oppressive to the individual, or very obvious to an ordinary observer,—I say, suppose such a case, and then it is, that the connexion between the general disturbance and divers local disorders is liable to be overlooked. It was, in pointing out this almost latent connexion; it was in demonstrating this important but neglected truth, that raised, and justly raised Mr. Abernethy to the highest eminence in his profession. His acuteness and penetration enabled him to trace the insidious march of diseased action from the imperfect digestion, and assimilation to the general contamination of the constitution, and thence to divers forms of local disorder in the different tissues and structures of the body; a principle of vast application in the practice both of physic and surgery—a principle, the elucidation of which will ever rank Mr. Abernethy amongst the most gifted and most successful cultivators of medical science. But, gentlemen, whilst I freely pay the tribute of my respect and admiration to this distinguished individual, I fear we must admit that this, like other human benefits, has not been without its alloy. The whole history of physic is a history of extremes, to which, indeed, there appears to be at all times and in all matters, a remarkable proneness in the human mind. Not only are we liable to be captivated by a newly discovered truth, not only are we liable to fall into excess in its application, but we are moreover extremely apt to doubt, neglect, or even reject our previous knowledge, merely because it happens to form a part of the creed of those who lived before us, and whom we are too much given to regard with complacent contempt. Mr. Abernethy's work on the *Constitutional Origin of Local Diseases*, I need not repeat, sheds a new light upon numerous diseases, previously ill understood; but, gentlemen, I cannot persuade myself, that it did not at the same time have the effect of diverting attention too much from the opposite of his position, from the consequences of *insidious local irritation* upon the *general constitution*. Proceeding with something little short of enthusiasm upon the newly established principle, men began to discover a purely constitutional origin for all local disorders; they forgot, or at least slighted the secondary effects produced upon the general system, by the continuance of local irritation, and regulated their practice accordingly. Of course I now speak of that form or species of local irritation, which proves injurious rather from its *continuance* than from its immediate *severity*; and it is the form of local irritation to which I am anxious in this lecture to solicit your earnest attention—an irritation often so slow and gradual in its approach, as to pass unheeded by the sufferer, and overlooked by the practitioner; an irritation nevertheless, which, whatever may have been its origin, shall by its continuance insidiously undermine the general constitution, and give rise to remote consequences subversive of almost every personal comfort—I mean then, gentlemen, *Uterine Irritation*.

I will not occupy your time and attention by any elaborate attempt to define the term *irritation*, a term in common use, to signify a disturbance in the endowments or functions of a part independent of either actual inflammation or organic lesion. It is not denied that either the one or the other of these states may occasionally attend or result from such irritation; all that is meant by the term is, as I have said, a disturbance in the endowments or functions of a part without either inflammation or organic lesion being *necessarily* present. In thus characterising uterine irritation therefore I do not deny that it may be coexistent with, or even give rise to inflammation or organic lesion; but speaking generally, I would say that it is found to exist altogether independently of either of these states.

The first question then is, how are we to ascertain its presence? What are the signs or symptoms by which it is to be recognised? A person of experience will almost infallibly pronounce it to be present, from a mere glance at the effects which it so commonly produces upon the system at large, and upon particular parts, and which usually prevail to a greater or less extent in every instance before a professional man is consulted: but, in addition to these, which I shall particularly point out by and by, there are certain signs or symptoms emanating more immediately from the uterus itself, some of which will, in a large majority of cases, enable

8

the practitioner to decide; that is, supposing him to make the inquiry, for unless he do so, and very keenly too, patients, either from delicacy or from ignorance of its importance, will most assuredly conceal them.

The most frequent symptoms of uterine irritation are, *irregular menstruation*, the discharge being preceded or accompanied by pain in the back, loins, or thighs, or in the region of the uterus itself, and attended with forcing or bearing down; the *discharge being in excess* either in point of mere quantity or in continuance, or in recurrence; *tenderness of the womb itself upon pressure* made either externally or *per vaginam*, a tenderness in some instances so great as to interfere with the privileges of matrimony; and, lastly, *leucorrhœa*. The most frequent symptoms, however, are, unquestionably, *painful menstruation* and *leucorrhœal discharge*, although the former is often the only symptom acknowledged by the patient herself. Such, gentlemen, are the few plain, simple indications of a state of uterus which is repeatedly overlooked, although productive of the most serious disturbance both of the general health and of particular organs; disturbance which, when once produced, stamps a character upon the general and local ailments of the sufferer, strongly indicative, to the experienced man, of uterine irritation; a character which confirms him in the belief that it is from such irritation that the evil originates, and that it is to correct the condition of the uterine system that his chief attention is to be directed.

Before proceeding, however, to notice in detail the constitutional and local effects of this continued irritation, it may not be improper to say a few words respecting the predisposing and exciting causes of the irritation itself.

The most powerful predisposing cause of uterine irritation is *constitutional irritability*, especially in persons naturally of a delicate frame of body; a state rendering its possessor acutely susceptible of impression generally, and which has not unfrequently, and certainly not very inaptly, been distinguished by the term nervous temperament. The rest of the predisposing causes are such as tend either to produce or to increase morbid susceptibility; such as *sedentary and luxurious habits, late hours* and *passions of the mind*. Natural irritability of the uterus itself may undoubtedly prove a predisposing cause, but of this it is difficult or impossible to speak with certainty, as the refinements and restraints of society will always furnish, at the same time, other co-operating causes.

The *exciting causes* again, are *active exertion of any kind during the flow of the menses; frequent child-bearing, especially if the patient suckle her children herself; excessive venery, and, indeed, venereal excitement of every kind.* Married women, I think, perhaps suffer most from child-bearing, and from imprudence during the menstrual period; unmarried women, on the other hand, from similar imprudence, and, peradventure, from causes of excitement of the genital organs, concerning which it is unnecessary to be very explicit.

I shall now proceed to describe to you, in the most intelligible manner I am able, the morbid conditions of the general system and of particular organs which result from the continuance of that uterine irritation to which I have just drawn your attention, only premising that they of course vary very much both in kind and degree, but chiefly in degree, according to differences in particular constitutions, and according to the susceptibility of individual organs.

The first complaint usually made by a female suffering from this irritation, is, of feeling *nervous*, probably with a disposition to be low-spirited; but if, in a well-marked case of the kind, you proceed to feel the pulse, and especially if it be your first visit, the chance is that you will perceive a distinct tremor of the hand, and a remarkable acceleration and sharpness of the pulse, evidently arising from mental agitation—a degree of mental agitation so great, in a naturally susceptible female, that if you attempt to soothe or encourage her, she will begin to sob, her lips quiver, and she bursts into a flood of tears. Even after she has sufficiently recovered her self-command, she experiences considerable difficulty in describing her feelings and sensations, and often appears to despair of satisfactorily communicating to you the nature of her ailment. She tells you, that, without any assignable cause, she gradually declined in health and spirits; that she has lost her wonted alacrity, has become indolent, and is easily fatigued by comparatively slight exertion; that she is readily flurried; that her heart often beats, flutters or palpitates; that the impressions made upon her mind are altogether disproportionate to the causes producing them; that she is very prone to weep, and occasionally experiences sudden and transitory feelings of alarm and dread, especially during the night, without being able satisfactorily to account for them; in short, that both body and mind are in a morbidly sensitive condition, whilst general distress is strikingly depicted in her pale or dejected countenance. If you proceed in your inquiries to ascertain the state of the uterus, you are perhaps informed that she is regular; but if interrogated more narrowly, it will almost uniformly be found that she suffers pain, either before or during the flow, in the back or loins, and that in the intervals she is troubled with leucorrhœal discharge. To these inquiries she gives a reluctant reply; will often, perhaps from delicacy, conceal the truth, or, if she acknowledge it, will probably add, "Oh, that is of no consequence; that is not my complaint; I have long been accustomed to that, and it does me no harm;" and then winds up the case with telling you that she has taken a load of tonic medicines without benefit. If you ask her whether such questions were ever put to her before, you are generally answered in the negative. Such a case as this, with slight modifications, is of common occurrence, yet it most frequently happens, that, with the general disorder, we have decided derangement of some internal organ or organs; and of these, the organs of digestion appear to be almost uniformly the first to participate; indeed the derangement of the digestive organs, to a greater or less extent, is so commonly associated with the general affection I have described, that one cannot but conclude that the general affection is most materially influenced, if not in part produced by it. It is sufficient, however, for our purpose to know that the exalted susceptibility of the general system and this deranged condition of the first passages, are very commonly coexistent, however they may stand in the relation of cause and effect.

The first appreciable disturbance of the stomach is most frequently a tendency to *flatulency*, which flatulency is productive of different effects in different individuals, although, in all, the stomach itself appears to be in a morbidly irritable condition, so as greatly to modify or aggravate the consequences that would otherwise arise from the presence of such flatulency. The patient experiences uneasiness at the scrobiculus cordis; she complains of a sense of load or distension after meals, or, if the stomach be uncharged with food, of prickings and anomalous pains in the organ, all of which symptoms are pretty uniformly relieved for a time by the expulsion of flatus from the stomach. In other cases, the irritation produced by the flatus about the cardiac orifice, excites a sympathetic affection in the throat, a sort of globus hystericus, which is variously described by patients, some calling it spasm, whilst others compare it to a mechanical obstruction, and indeed one lady somewhat fancifully compared it to a bullock in her throat. It is a sensation, however, which will often last, in a greater or less degree, for days, or even weeks, with little intermission.

At other times the patient suffers from repeated vomiting, or is perhaps seized suddenly, but only occasionally, with vomiting, preceded, accompanied, or followed by an irregularly inverted action, chiefly of the œsophagus, and attended with an ascent of flatus, so as, in some instances, to threaten suffocation.

Such, gentlemen, are the ordinary affections of the stomach met with in the disorder to which I have alluded; affections, however, varying both in kind and degree in different individuals; affections, too, in which the whole alimentary canal appears more or less to participate, the patient being very commonly troubled with rumblings, distension and anomalous transitory twitchings in the bowels, symptoms undoubtedly depending upon the flatus and other contents irritating these already morbidly sensitive organs. In more rare cases there appears to be a sort of inverted action of a greater or less portion of the alimentary canal; an inverted action commencing in some part of the bowels; an inverted action accompanied by a rumbling noise in these organs; an inverted action extending to the stomach, passing up the œsophagus to the pharynx, where it produces a most distressing sense of suffocation, and lastly communicating with and involving in one universal disturbance the brain and entire nervous system. This mysterious communication so closely allied to the aura epileptica is in all probability somewhat of the same nature, and may, perhaps, result from the irritation in the alimentary canal being conveyed through the pneumogastric nerve to the brain; for when this takes place, the breathing is usually affected at the same time in a very remarkable manner;—the whole assemblage of symptoms, the aberration of mind and bodily contortions constituting, when taken collectively, what in common language is understood by an hysteric paroxysm. There are, however, divers modifications met with—the patient being sometimes seized with violent and involuntary fits of laughter or crying during the paroxysm, at other times she will lie in a perfectly motionless and insensible state, but with a natural pulse, for hours together; or she shall suffer from a more or less perfect paroxysm without its being preceded by any globus whatever.

Hitherto then, we have traced the effects of uterine irritation as they appear in the form of a general morbid sensibility and irritability of the stomach and bowels; which symptoms alone, but in different degrees, may continue to harass such patients for months, or even years, without either a well-developed hysterical paroxysm, or any other inconvenience deserving particular notice. But it remains to be observed, that this exalted state of the nervous system is not, in a considerable proportion of cases, evinced solely by morbid susceptibility and excessive or irregular action, but, moreover, by *morbid sensation*, affecting different parts or organs of the body, and this too varying from the slightest uneasiness to the most exquisite torture.

Now, gentlemen, it is to these painful affections that I am anxious to direct your earnest and special attention; because, in the first place, it is to relieve these that you will most frequently be called upon to render your assistance, and because, in the next place, these painful affections are perpetually misunderstood and mistaken, to the serious detriment of the patient. I repeat then, I am anxious, most anxious, to impress upon your minds the nature of these secondary painful affections, their most frequent seat, and the various sources of fallacy by which you may be betrayed into error.

Of these painful affections, the most serious, or at least the most prominent, and certainly by far the most interesting, are those which attack the abdominal viscera, as these are repeatedly mistaken for inflammation, and treated accordingly. I shall not stop to inquire why those viscera, supplied by the ganglionic system in general, or why the abdominal viscera in particular, should be more liable to suffer from such painful affections than other parts of the body supplied by the ordinary nerves of sensation and voluntary motion; suffice it to say that such is the fact.

Of the painful affections of the abdominal viscera the most frequent are—

1st. A PAIN SEATED UNDER THE LEFT MAMMA, OR UNDER THE MARGIN OF THE RIBS OF THE SAME SIDE.

2ndly. A PAIN UNDER THE MARGIN OF THE RIBS OF THE RIGHT SIDE.

3rdly. PAIN IN THE COURSE OF THE DESCENDING COLON.

4thly. PAIN IN THE COURSE OF THE ASCENDING COLON, ESPECIALLY TOWARDS THE RIGHT HYPOCHONDRIUM.

5thly. PAIN AFFECTING THE ABDOMEN GENERALLY.

6thly. PAIN IN THE REGION OF THE STOMACH.

And lastly, PAIN IN THE REGION OF THE KIDNEYS, SOMETIMES EXTENDING DOWN THE COURSE OF THE URETERS TO THE BLADDER.

Such are the painful affections usually met with, attacking the abdominal viscera; affections too, which, in point of frequency of occurrence, observe the order in which I have enumerated them. You will have noticed that I am reserved in referring these pains to any particular organ, contenting myself with the *fact* that the pain is felt in a particular *situation*. The truth is, considerable difficulty presents itself in ascertaining positively what particular organ is affected, as the natural functions of the part or organ do not, in such cases, appear to be necessarily disordered, or at least to such an extent as to indicate with certainty that it is the seat of the pain complained of by the patient; neither can we have the usual aid of *post-mortem* examinations, as few die of such complaints, whilst, if the patient be cut off by another disease, no organic lesion is discoverable to satisfy our inquiries. Occasionally, however, this question may be decided during life, as we shall see presently.

With respect to the *pain under the mamma, or under the margin of the ribs of the left side,* this is out of all proportion of most frequent occurrence, and will often last for weeks, or even months together, with but little intermission. This pain is very circumscribed; it is not necessarily or constantly increased by a deep inspiration or by external pressure, although this is occasionally observed; it is seldom attended with cough; it is not materially affected either by a charged or by an empty state of the stomach, but varies in its intensity, and now and then goes off altogether for a few minutes, hours, or even days, or the pain shall subside and be succeeded by a mere uneasiness or sense of fulness in the part. This pain, as I have said, is of extremely frequent occurrence, and is very often associated with palpitation of the heart, or, what is much more usual, with unnatural pulsation of the organ, if I may be allowed the expression; *i. e.* the patient is conscious of the heart's action, or she feels as if its impulse were communicated to a part so sensitive as to excite distinct sensation, which, you know, is not the case in a state of health. With respect to the precise source of the pain, I confess myself at a loss to speak with confidence or certainty, but am upon the whole inclined to assign it, when complained of under the left mamma, to the cardiac orifice of the stomach; at least in one case in which it had prevailed for a considerable period, and in a very aggravated degree, I was led to this conclusion. The young woman, to whose case I allude, died suddenly in a fit, and I examined the colon, spleen, heart, and stomach, with the minutest attention, when the only indication of irritation I could detect was a ring of very delicate vessels, or rather a blush of redness surrounding the cardiac orifice of the stomach, such as might be supposed to be the result of any continued irritation or spasmodic action. Whatever may be its precise seat, it is repeatedly, but erroneously, supposed to be purely of an inflammatory nature, and consequently is mistaken for and treated as pleuritis or splenitis.

The second painful affection to be noticed is that *seated close to the margin of the ribs on the right side*. This pain, gentlemen, although occasionally circumscribed almost to a point, usually extends from the scrobiculus cordis along the margin of the ribs, nearly to the loin of the side affected; it is neither considerably nor uniformly increased by a full inspiration, yet this is occasionally observed. External pressure, however, aggravates the pain, and sometimes in a very remarkable degree, whilst in some instances there is such tenderness, that the patient shrinks from the slightest touch. The pain now and then shoots through to the back or between the shoulder-blades, but very rarely to the top of the right shoulder. The pain will occasionally remain, with slight remissions, for weeks or even months; at other times, it subsides altogether, and is succeeded, like the pain under the left breast, by a sense of fulness or tension of the part. As to the actual seat of this pain, gentlemen, I again confess myself incompetent to decide. I have sometimes supposed it to be in the *colon*, as it may now and then be traced from the margin of the ribs into the right iliac region; in other instances I have supposed it to be seated in the *duodenum*, from its being occasionally attended with sickness, from its being aggravated during the operation of mercurial purgatives, and from its being in some rare cases attended with a remarkable sallowness or icteritious aspect of the countenance, and indeed with almost complete jaundice.

Here again, then, I must leave you in doubt, merely remarking, that it is not inflammatory, although repeatedly mistaken for hepatitis, and treated as such.

The next painful affection to be noticed is *that seated in the course of the descending colon*. This is not unfrequently associated with the pain under the left mamma, but is also observed to exist alone, extending from below the ribs to the *sigmoid flexure of the colon*,

This, like the others described, is variable in its degree, and although more or less permanent, sometimes remits for hours, days, or even weeks together, and again returns. It is, however, more decidedly and obviously effected by flatulence, than the pains in the other situations mentioned; thereby more clearly pointing out the seat of the malady. The movement of the flatus is occasionally attended with a gurgling or rumbling noise, and a simultaneous aggravation of the local pain in the part. This is now and then mistaken for colitis, or for some organic lesion of the organ.

Fourthly, *Pain in the course of the ascending colon.* I have had occasion to observe, that the pain already described as situated behind the margin of the ribs on the right side, sometimes extends down the course of the colon as far as the iliac region; it not unfrequently happens, however, that the pain is felt exclusively in the situation of the ascending colon, and like that on the opposite side, varies in degree at different times, or for a period disappears altogether. This pain too, after a time, is attended with considerable tenderness, so that the least pressure creates inconvenience.

Fifthly, *Pain affecting the abdomen generally.* This is by no means of rare occurrence, and in some instances so closely resembles general peritonitis, as to be mistaken for, and treated as that complaint. Indeed, I know of no disease more puzzling than this, and it was not, I confess, till I had witnessed several such cases, and attended minutely to the history and progress of the disorder, that I became convinced of its true nature. It may be called a general *neuralgia* of the abdomen. It is sometimes attended with a tympanitic, and at other times, with a flaccid state of the bowels, the former being by far the most distressing. The pain is complained of over the whole of the belly, and the slightest touch, in many instances, cannot be borne, such is the extreme sensibility and tenderness of the parts.

If you watch the case attentively, you will, in general, soon detect some incongruity in the symptoms, to excite doubt and suspicion; but yet, so close is the resemblance in some cases, as almost to set positive diagnosis at defiance. A case of this kind lately occurred in Martha's ward, when I was in so much doubt, that, to err on the right side, I treated it as peritonitis, although the history of the patient, and the condition of the uterus, told against it: I soon became convinced that I had been wrong, as I shall presently explain to you. It may be observed too, that this general pain of the belly, like peritonitis, frequently occasions but little distress, unless pressure be applied, and, like peritonitis also, it suffers an aggravation at intervals, an aggravation apparently in both cases depending upon the spasmodic action or griping in the bowels.

The *sixth painful affection* is that attacking the *stomach* in particular. This pain is for the most part strongly marked, and the more intense the disorder, the more positive is the evidence of its being really seated in the organ mentioned. Thus it will sometimes come on suddenly, occasioning the most excruciating agony; the patient screams from the violence of her sufferings, her countenance is expressive of the greatest distress, and she leans forward or bends the body, in order to diminish the pressure of the abdominal parietes; or, she says that the pain is drawing her double. This, in some cases, will last with little mitigation, for several minutes, or even hours, the patient the whole of the time making loud complaints, and declaring that she must die if not speedily relieved. This pain will probably remit, and be succeeded by another much less severe, though more permanent; both the intense and more moderate pain, being much increased by pressure made upon the epigastrium.

Lastly.—Such patients occasionally experience a sudden and severe attack of pain in the region of the kidneys, to which region it may be exclusively confined, till it disappear altogether, or it shall extend from thence down the course of the ureters to the bladder, or the bladder alone shall be affected with pain; in either of the latter cases the patient generally experiences more or less dysuria.

Having pointed out to your notice the ordinary local affections arising from or connected with uterine irritation, I need only add, that you will often in the same individual meet with more than one of them existing at the same time; or, what is more usual, you will have them alternating with each other; whilst you will also observe the greatest variety in the relative severity of the general and local disorders mentioned; the general derangement prevailing occasionally in a very high degree with but little local pain, and *vice versâ*.

Another, and most important question is, the *Diagnosis*, or the means of distinguishing the painful or neuralgic affections I have described from the inflammatory disorders to which they often bear so close a resemblance, and for which they are so repeatedly mistaken. This, it must be confessed, is not only important, but unfortunately sometimes as difficult as important. I flatter myself, however, that you will be less likely to be misled now that your attention is alive to the sources of fallacy—sources of fallacy, gentlemen, which, in common conversation, all admit and profess to know; a source of fallacy, nevertheless, which is more freely acknowledged than attended to in practice, otherwise we should have fewer errors committed than, I know too well, are committed even at the present day. Knowing then that such cases are of frequent occurrence, whenever a female complains to you of pain under the left breast, with or without palpitation or pulsation of the heart; of pain in the right hypochondrium; in the situation of the left or right colon; or of acute pain generally over the whole belly, or in the region of the kidneys or bladder—always be upon your guard, and if on inquiry you find a few, or many of the constitutional symptoms I have described, together with indications of uterine irritation, as shown by pain in the pelvis, in the loins, or in the thighs, before or during the catamenial flow; by too frequent or too profuse menstruation; or by leucorrhœal discharge; I say, when you find such an assemblage of symptoms and circumstances, your suspicions will amount to a high degree of probability, that the complaint is not of an inflammatory nature. The comparative absence too of fever, the appearance of the tongue, and the state of the surface, will materially assist you in the diagnosis; although it must be confessed that peritoneal inflammation often exists with but little development of the general symptoms regarded as characteristic of the phlegmasiæ; whilst on the other hand, the painful affections I have been describing are not unfrequently attended with an accelerated, and vibrating, or even a hard pulse, a hot skin, and furred and slightly brown tongue. Generally speaking, however, the pulse *varies much* in the latter, and seldom has much hardness in its beat, although from the irritability of the heart, it is sometimes sharp or jerking, whilst the skin is usually *warm*, rather than *hot*, and with a disposition to moisture; and lastly, the absence of many, or of some of the symptoms of the phlegmasiæ simulated, will go far to complete the marks of distinction. But more of this anon.

It must, nevertheless, be carefully remembered that all or most of the symptoms I have enumerated, as emanating more immediately from the uterus itself, may and often do result from organic changes either in the uterus or in the parts adjacent. Whenever therefore we have the least suspicion of the existence of such causes, an examination of the patient ought on no account whatever to be neglected.

The necessity of such precaution may be said to increase with the age of the patient; but when obstinate or repeated flooding prevails, whether slight or profuse, such an examination becomes quite imperative, especially when it occurs in females somewhat advanced in life.

I shall not occupy your time by entering upon any minute detail of other morbid affections occasionally arising from the same source. Those enumerated are of most common occurrence, and are most frequently mistaken and maltreated. Amongst the more anomalous and more rare disorders of the kind, I may merely mention the *irritable mamma* and *obstinate pain in the head*, both of which I believe, occasionally, at least, to depend upon the state of uterus alluded to; whilst many local physical disorders are materially influenced by it, a circumstance which has not escaped the observation of my friend Mr. B. Cooper. Neither will time permit me to point out those peculiarities which present themselves in chlorotic girls, and those observed at the advanced period of life, when the functions of the uterus are about to cease. To explain these peculiarities and the differences of treatment founded thereon, would be incompatible with the limits of a clinical lecture; I shall, therefore, proceed to the *treatment* and the mention of a few of our cases in illustration.

The *indications* are, 1st. *To correct the morbid condition of the uterus*. 2ndly. *To remove or mitigate the violence of troublesome symptoms* in any individual case; and 3rdly. *To restore tone and vigour to the general constitution*.

In thus laying down the indications of cure, you will perceive that I have ventured to reverse, in some measure, the ordinary mode of procedure. I repeat to you, gentlemen, that the condition of the uterus, to the continuance of which, I ascribe the various disorders enumerated in this lecture, has in most instances been altogether overlooked, or at least, if observed, it has only been in those cases which have had some of the characteristic symptoms present, in a strongly marked degree, such as excessive or very painful menstruation, or profuse leucorrhœal discharge; and even then, the condition of the uterus has been usually looked upon as a mere effect or consequence of the nervous or hysterical state of the entire frame, or in other words it has been considered as merely participating in a universally morbid sensibility. Hence the primary indication has commonly been to restore strength to the general system, regardless of the local disorder, although the ill success of the treatment founded thereon has at all times been frankly acknowledged; so much so, indeed, as almost to have placed the disorders in question amongst the *opprobria medicinæ*. Having been led however to take a different view of the subject, I am necessarily induced to make the correction of the state of the uterus, the first, or at least, the principal indication, although of course, the second indication, that of allaying troublesome symptoms, must often go hand in hand with, or in aggravated cases even take precedence of it. What then are the means best calculated to remove the morbid condition of the uterus, the irritable state of the organ? In the cases in which the uterine disorder has attracted attention, it has been recommended to bleed generally, to take blood by cupping from the loins, or by leeches from the region of the pubes, or from the pudendum; to purge, to give anodynes, to employ the warm bath, and to keep the patient in a recumbent position, even for months at a time. Of this kind are the principal remedies recommended in an excellent work recently published by a distinguished member of our profession—I mean Dr. Gooch. That gentleman treats of the complaint as it is attended with actual tenderness of the os uteri to the touch, when an examination is made *per vaginam*. He enumerates the remedies I have mentioned, but with those expressions of extreme distrust, which mark the man of candour and of truth, and with that modesty which, when united to extensive experience, commands the assent of every reader. Now, gentlemen, that all these remedies may be occasionally serviceable, or even necessary, I am not presumptuous enough or inclined to deny: depletion, general or local, in the plethoric, and anodynes and laxatives in most cases; yet speaking from experience, this object, this grand object is secured with much greater certainty, and much more speedily, by applications made directly to the uterus itself, and parts adjacent. The applications to which I allude are *cold astringent washes*, injected *per vaginam* by means of a proper syringe. The ordinary womb syringe answers the purpose exceedingly well, but one of any convenient shape may be used, provided it be sufficiently large to contain from four to six or eight ounces of fluid. The injection should be introduced with such a degree of force, as shall secure its application to the upper part of the vagina, and to the os uteri; and the operation should be repeated two, three, or four times a day, according to the circumstances of the individual case.

Either the mineral or vegetable astringents may be used; the former however I prefer, as they do not stain the patient's linen, and consequently are not so much objected to. With respect to the precautions to be observed in the employment of these injections, very few are required beyond what common-sense would dictate. Should the injection occasion smarting, which is by no means unfrequently the case at first, it may be diluted with water, or water alone may be used till the original tenderness subsides, which for the most part it will soon do. It will also be prudent to instruct the patient to relinquish it a little before the expected period of menstruation, and to resume it as soon as that period is over. These are almost the only precautions I have ever deemed it necessary to observe. Although in very irritable habits, and especially when the stomach is liable to be affected with pain and spasm, it may be as well to direct the wash to be used tepid at first, gradually diminishing the warmth till it is brought to the ordinary temperature of the patient's apartment, which will pretty uniformly be borne exceedingly well after a time, except perhaps during a few of the coldest months in winter. The wash I most frequently employ is the *Liquor Aluminis Compositus*, of the London Pharmacopœia, *i. e.* two drachms of *alum*, and two drachms of *sulphate of zinc*, to a pint of water. This practice must be persevered in for a length of time, proportionate to the obstinacy of the case and the effects it produces. Indeed, I myself recommend females never to relinquish it, but to employ it from time to time, as long as they continue to menstruate, to prevent the recurrence of the disorder, and its unhappy consequences. I have said that the patient should desist from the use of the injection a little before and during the menstrual period, but she ought also to be specially cautioned against using any violent exertion, or undergoing any unusual fatigue at that time, as nothing so completely thwarts your purpose as imprudence committed whilst the irritable uterus is performing its functions.

When the uterine irritation is characterised by frequent and excessive flow of the menses, I have directed the patient to remain quiet in or on the bed, and to desist from the washing during each recurrence; not having ventured to carry the practice to the extent of attempting to restrain even such excessive discharge by local astringents. But when there is decidedly painful menstruation, or pain felt in the *womb*, *loins*, or *thighs* before the appearance, although there be no leucorrhœal discharge whatever in the interval, I nevertheless apply the cold wash to such subjects, precisely in the same manner as if such leucorrhœal discharge were present, on the principle that this leucorrhœal discharge is itself a mere symptom or effect of the state of the uterus and neighbouring parts, which I am anxious to remove. Such, gentlemen, is the *local* treatment that experience and observation have led me to adopt, and which, after a long trial, I venture, with some confidence, to recommend to you. It is just possible that circumstances not noticed by me may occasionally occur to interfere with or forbid the practice; but, gentlemen, I speak of generalities, and not of universal principles—universal principles are unknown in our profession.

I would only remark further, that, very recently, Mr. Jewell has recommended the Nitrate of Silver, either in substance or in solution, as an application in leucorrhœa; and I have no doubt, from what we observe in its general use, that it will have a powerful effect in destroying morbid sensibility; but I have had no experience with it myself, whereas I have extensively employed the astringent wash before mentioned for the last ten years, and declare to you that I have never known any serious inconvenience to result from it in a single instance; if ever injurious, therefore, it is unknown to me.

The second indication—*to allay or remove troublesome symptoms*—must, of course, be variously fulfilled, according to the circumstances of the particular case, but the remedies employed for the purpose will necessarily co-operate with the local treatment to afford relief to the patient; unless, indeed, the severity and character of the troublesome symptoms be such as to render the *immediate* application of the wash either doubtful or hazardous—as *e. g.* when the stomach is the seat of severe spasm, or the bladder or abdomen generally acutely painful, in which cases the second indication must take precedence of the first. Should the symptoms merely consist of the susceptible state of body and mind before described, or should this be accompanied by that modification of derangement of the digestive organs characterised by flatulency and irritability of the stomach and bowels, the general treatment need be very simple. I say nothing of *bloodletting*, as the propriety or impropriety of this must be apparent from the state of the circulation, always keeping in mind that such subjects bear bleeding badly, especially in large quantity, and that mere *frequency* even with *sharpness* of the pulse, is, in the highest degree, fallacious: a full, hard pulse, with considerable heat of skin, and more or less throbbing or pain in the head, will point out the expediency of a moderate bleeding. *Purgatives* or rather *Laxatives*, ought never to be neglected, but much caution and judgment are required in their proper selection and application. The bowels, it is true, are generally costive, but it must be remembered that they are as uniformly in a weak and irritable state; hence those purgatives should be selected which give the least pain, and have the least tendency still further to weaken their tone. Watery saline purgatives are ill-suited to the majority of such cases; they often perform their office imperfectly, and although they may not produce much pain, which they often do in such subjects, they, nevertheless, tend, by frequent use, further to impair the tone of the bowels, and to increase the disposition to flatulency. Except, therefore, at the commencement, in plethoric subjects, or where the disorder exists in connection with menorrhagia, the saline purgatives are not commendable. *Castor oil*, when it agrees with the patient's stomach, will be found more generally serviceable, perhaps, than any other. In many cases, the warm and resinous purgatives, though at all times more irritating, answer exceedingly well, emptying the bowels of their feculent contents, and expelling flatus without leaving any increased tendency to flatulency afterwards, as is the case with the watery saline purges. A little *Compound Extract of Colocynth*, or equal parts of this and *Extract of Rhubarb* with a little *Extract of Hyoscyamus* to obviate griping; or the above extracts may be given with a little *Blue Pill*, should a mercurial laxative be indicated by the appearance of the secretions. Calomel very often gripes severely, whether alone or combined; yet such is the variety of constitutions met with, that the Compound Extract of Colocynth with Calomel often proves the most effectual, and not the most painful purge. One or other, then, of these laxatives may be necessary once or twice a week.

In the case I am now treating of, that is, where there is no abdominal pain, the medicine that I have found to afford the greatest relief is unquestionably the *Ammonia*, given either in common mint or camphor julep, alone or with a few grains of *Magnesia ;* about ten or fifteen minims of the *Liquor Ammoniæ Subcarbonatis*, with from eight to ten or fifteen grains of *Subcarbonate of Magnesia* two or three times a day. If it create any approach to pain in the stomach, or a disagreeable sense of heat there, these effects may sometimes be obviated by adding about half a drachm of Tincture of Hyoscyamus, or a drachm or so of the Tincture of Hops, to each dose. This ammoniacal mixture seems, by its stimulus, to expel flatus, and thereby to afford relief to many of the uneasy and unhappy sensations, and to raise the spirits of the patient. I have often given the Ammonia along with the Mistura Myrrhæ of Guy's Hospital, an ounce or a dose of which contains, I believe, about twenty grains of Myrrh, and is made with the Decoction of Liquorice root.

But suppose that, with more or less of the general symptoms, we have pain under the left mamma, or in any of the other situations pointed out, what modification of treatment ought we then to adopt?

The pain under the left mamma is of such frequent occurrence, that it becomes a matter of the very first practical importance to bear it carefully in mind when called upon to treat the disorders of young females. There is no local pain more frequently mistaken, and there is, perhaps, no local disorder so maltreated as this. Over and over again have I known patients blooded, cupped, blistered, and anointed with acrid ointments, for months in succession, not only without relief, but with the most serious injury to their general health, whilst the uterine irritation has been altogether overlooked. Indeed, gentlemen, in thus declaring and denouncing such a mistake and such bad practice, I candidly confess that I am only declaring and denouncing "*quæque ipse miserrima vidi, et quorum pars magna fui.*" In truth, it was the frequent occurrence of this obstinate and intractable pain that first led me to a close and minute investigation of the subject of uterine irritation, having, like others, repeatedly blooded, cupped, and blistered in vain in such cases. Since, however, I have been guided in the treatment by the principles I have ventured to lay before you, my success has been much more certain and decided, and when the treatment founded thereon has not fully answered my expectation, it has at least been attended with the negative advantage of inflicting no unnecessary injury on the patient's general health.

In pointing out its non-inflammatory nature, gentlemen, do not imagine that I contend for the impropriety of bleeding or cupping, or blistering in every instance. On the contrary, in plethoric subjects, one or two general or local bleedings may occasionally prove of service, but merely, I believe, as such practice might be expected to afford relief in an ordinary colic, or any other similar painful affection. Depletion, however, is very far from being in general necessary, and probably will, in a majority of cases, be positively hurtful,

9

teasing and exhausting the patient, without any adequate or permanent benefit. I have known Cupping, Leeches, a Blister, or an Opiate or Belladonna plaister afford relief, but they all often fail, and will generally do so, or at least be merely followed by a temporary respite, unless the condition of the uterus, and other circumstances I have pointed out, be attended to. Hot fomentations to the part will sometimes afford relief, but are attended with the same uncertainty as the other local remedies mentioned. Under the use of the injection, however, and the ammoniacal mixture, with Tincture of Hyoscyamus, or with three, four, or five grains of the Extract of Conium, in the form of pill, alone or with a grain or so of blue pill, according to the state of the secretions, the pain will generally yield, although the patient, as might be expected, will for a longer or shorter period be liable to a recurrence, until the original irritation, and its effects upon the abdominal viscera, shall have been overcome by a steady perseverance in the use of appropriate remedies. In some instances the ammonia cannot be borne, when the plain Camphor, or Mint julep, may be substituted, with half a drachm or so of the Tincture of Hyoscyamus, for four or five grains of the Extract of Conium may be given in the form of a pill twice or thrice a day. In other cases again, you may try six or eight drops of Batley's Liquor Opii Sedativus at bed-time, unless it offend the stomach or the head. Should it constipate the bowels, proper laxatives must be had recourse to, in order to obviate such an effect; I prefer, myself, the Conium, or Hyoscyamus. Unless, however, it happen thus to disagree with the patient, the ammonia will answer best, guarded, of course, by the anodynes.

In those cases, in which the pain shifts its seat from time to time, I limit my *external* remedies generally, to hot anodyne fomentations; such as the Decoction of Poppy-heads, or what often answers very well, the hot Infusion of Chamomile Flowers, the flowers themselves being inclosed in folds of flannel, the more effectually to retain the warmth. I may also here observe again, that this tendency to shift its seat, together with the unsteadiness in the degree of the pain, form the most important diagnostic indications; whilst in most instances, we shall find, on inquiry, that the patient had suffered more or less from the complaint for some time before she applied for advice. Attention to these circumstances, will materially assist us in our diagnosis, not only in these, but in the other painful or neuralgic affections attacking the abdominal viscera.

When the pain is situated in the *right hypochondrium*, or under the margin of the ribs of the right side, it is, as I have before said, repeatedly mistaken for, and treated as, hepatitis, and many are the times that I have known poor delicate women blooded, cupped, blistered, and salivated on this supposition, till they have been brought almost to the very brink of the grave. It is to be distinguished by the symptoms and circumstances already pointed out, and when ascertained, is to be treated upon precisely similar principles to those so minutely detailed. In these cases, patients sometimes tell us that our medicines, especially mercurial and resinous purgatives, greatly aggravate the pain of the part during their operation. Cupping, Leeching, and Blistering, will each occasionally, but only occasionally, afford relief: whilst, generally speaking, I would say, that in this form of the complaint, hot anodyne fomentations, assiduously applied, are more decidedly beneficial than in the former case, but in other respects the same means may be pursued. In some cases, four or five grains of Dover's powder, twice a day, will afford relief; in other instances, the more powerful combination of calomel and opium has been tried with effect. The calomel, however, ought to be given with great reserve, and never, if possible, to such an extent as to affect the system; in short, in all cases, the less violent measures we employ the better.

When the pain seizes the track of the colon on either side, the same treatment will apply, as it will also when the stomach becomes the seat of the pain, unless indeed severe spasm accompany it, when more active measures must be had recourse to; as, a full dose of Liquor Opii Sedativus, or even a full dose of *Laudanum* and *Sulphuric Æther*, as I shall presently show.

I have said that the pain occasionally attacks the whole of the belly, exactly simulating acute peritonitis. This may be attended either with a flaccid or a tympanitic state of the intestines, the latter proving by far the most painful of the two, the slightest touch causing the patient to cry out.

This generally diffused pain is not unfrequently associated with that modification of uterine irritation marked by excessive menstruation, both in point of quantity and frequency. But in such cases, whether with or without menorrhagia, it will require all your tact and discernment, and, what is more, it will require all your philosophy and forbearance, to abstain from copious depletion under an apprehension that it may be peritonitis. This history of the case will generally raise a doubt, and will often bring conviction; but if a doubt do really exist, it will always be prudent to err on the right side, and treat it as inflammation. I shall mention a case or two of this kind presently.

When *unaccompanied* by excessive menstruation, the same general rules of treatment are to be observed as have already been detailed; moderate bleeding if much plethora and headache, with a full pulse, prevail; and the free use of laxatives and fomentations. In both forms of the complaint to relieve the pain in this situation, I hold the Liquor Opii Sedativus to be by far the most efficient remedy, as well as attended with the least inconvenience to the patient. With due attention to the bowels, therefore, six, eight, or ten drops may be given every night, or every night and morning, according to the urgency of the pain.

When attended with excessive menstruation, there very usually prevails a plethoric condition of such irritable subjects, so that a moderate bleeding will commonly form a very good mode of commencing our practice: or if there be much forcing, or bearing down, the patient may be cupped from the loins. In all cases of this kind too, the patient during the flow, and for a short time before the expected recurrence, should be strictly enjoined to remain in the horizontal posture, to be kept cool, and to live upon low and bland diet. Nothing so certainly, or so greatly aggravates cases of menorrhagia, even when without abdominal neuralgia, as the ill-founded, and ill-judged practice of giving tonics and stimulants, on the erroneous supposition that such excessive discharges are the result of weakness. Repeatedly have I known the most serious mischief result from the practice here pointed out; patients having been given Bark, Port wine, and cold Porter, merely because they complained of great debility, and because such complaints of debility have tallied with the preconceived notion of the practitioner, that the disease had its origin in weakness. You must be careful, gentlemen, to distinguish a *weak and relaxed*, from a merely *irritable*, or a *delicate* and *irritable* habit: it is the latter in which you most frequently meet with the menorrhagia, of which I now speak. The excessive discharge is intimately connected with such irritability of the system at large, and of the uterus in particular; tonics and stimulants only tend, therefore, to produce excitement; to augment the action going on in the irritable uterus, and thereby to increase the sufferings, and exhaust the strength of the patient. As far as the discharge is concerned, therefore, your object is to allay the vascular excitement by moderate depletion and purging, and to allay irritation by anodynes and rest. A free action on the bowels exerts so powerful an influence over excessive uterine discharge, that I am often induced to administer divided doses, either of the Magnesia and Salts mixture, or of the Salts dissolved in the Compound Infusion of Roses; endeavouring to obviate the increase of the neuralgic pain of the belly, which is liable, when it prevails, to be aggravated by the Salts, by giving at the same time about five grains of Extract of Conium, or Extract of Hyoscyamus, three times a day instead of the Liquor Opii Sedativus. In general, however, other laxatives may be given occasionally, to maintain a free discharge from the bowels, and to counteract the effect of opiates, should we determine upon a trial of them; for this purpose Castor-oil, or Calomel and Rhubarb may be given: or the object may be, to a certain extent, accomplished by the use of glysters. If in spite of our depletion, laxatives, and rest, the flooding still continues to prevail, without neuralgia of the abdomen, I generally direct the patient to apply cold vinegar and water, by means of a sponge, to the lower belly, two or three times a day; employing at the same time anodynes, or anodyne glysters, in order to allay the irritability of the womb itself.

If you have flooding alone then, without neuralgia of the belly, you will bleed generally or locally, according to the state of the patient's system at the time; keep up a free action on the bowels; keep your patient cool, and on low diet, and administer anodynes, either by the mouth or by glyster, with or without cold sponging according to the severity of the case. Should flooding prevail along with the neuralgia of the belly, precisely the same remedies, with the exception perhaps of the cold sponging, will be required; whilst, if the neuralgia exist alone, the case becomes much more simple, must be treated on common principles; *i. e.* general bleeding, according to the degree of excitement of the heart and arteries: perhaps leeches to the belly, hot fomentations, laxatives, and anodynes, either given by the mouth, such as Liquor Opii Sedativus, Conium, or Hyoscyamus; or in the form of glyster, as thirty or forty minims of Laudanum, with a little thin starch. Even in flooding cases I never give the lead, as I never fail by more gentle means to subdue the disorder, and I have known the lead induce severe colic.

I need only observe further, that, if you should feel disposed to try the Oleum Terebinthinæ, I should consider it best suited to the general neuralgic state of the belly, without flooding or excessive menstruation in any form. From two drams to half an ounce may be given with three or four drams of castor oil. I am, however, averse to it myself from its liability to offend the stomach, and be rejected, but more especially from its tendency to act with violence on the kidneys. You had a good illustration of this in a female patient, No. 10, in the clinical ward; obstinate and dangerous Hæmaturia having been produced by Turpentine taken with a wicked intent.

Such, gentlemen, are the means which I have found most frequently of service in alleviating the more prominent and distressing symptoms attendant upon uterine irritation; means, I confess, often insufficient for the purpose, unless long persevered in, and this too in conjunction with the local treatment already mentioned. Many other medicines and expedients will undoubtedly suggest themselves in particular cases, as Camphor, Musk, the Warm-bath, and such like; or should the secondary local pains described, not be considerable, the *fetid gums*, especially *Assafœtida*, will often afford some, though tardy relief. Indeed these fetid gums have attained a high character as *antihysterics*, but appear to me, chiefly, if not exclusively adapted to relieve some of the symptoms more particularly connected with the condition of the bowels;—they expel flatus, raise the spirits, and moreover appear to restore tone to the alimentary canal, provided the original source of the evil be attended to at the same time. In the less painful forms therefore of the disorder, these fetid gums are worth a trial, and will often afford considerable relief to the feelings of the patient.

Much of what has been said, gentlemen, merges in the *third indication*; viz. to restore strength and vigour to the general habit. Of the best means of accomplishing this I need say very little, beyond remarking that the *early* use of *tonics* has been extensively adopted, and has consequently been given a fair trial, whilst almost all are agreed as to the very unsatisfactory results, in a large majority of cases; the cause of which you will now be able, in some degree, to appreciate; I mean, of course, inattention to the local irritation, from which I suppose the whole mischief to proceed. In the highly irritable and susceptible state of the body at large, and of the alimentary canal in particular, the more stimulating or irritating tonics cannot be borne, creating sickness, pain or uneasiness at stomach, loss of appetite, headache, and divers unhappy sensations not easily defined by the patient. Chalybeates not only offend in this way, but are extremely apt to excite the uterus, so as to produce excessive menstruation, if it did not previously exist, or to aggravate it, if it have already prevailed. Exceptions do undoubtedly occur, but the objections to them stated are founded on personal experience. The *Sulphate of Zinc*, however, may sometimes be given early, with advantage, provided it do not offend the stomach. A grain to begin with may be given night and morning, alone, or with a few grains of the Extract of Conium, or of Hyoscyamus, or with a little of the Pil. Galban. Comp.; or it may be given along with two or three grains of the *Extract of Gentian*, or *Extract of Bark;* or these latter may be given without the Zinc. Any of these may be tried, and persevered in, till the local disorder and the most prominent of the general symptoms shall have been subdued to a certain extent, when less reserve may be observed as to the nature of the tonics we employ; the chief of them are *Bark, Bitters, Chalybeates,* the *Cold Bath,* and *Country air.*

The prophylactic measures have already been alluded to, and need not be repeated, further than to impress upon you the necessity of the patients avoiding all irritation or excitement during the menstrual periods.

A most important matter to be attended to in every case, is the proper regulation of the diet of our patient. Bearing in mind that the stomach is almost uniformly in a weak and irritable condition, all those articles which are of difficult digestion, and all such as are calculated to produce excitement, ought to be most carefully avoided. Hence the patient should abstain from salted meats, made dishes, pickles, and high-seasoned food, whilst almost all vegetables, especially in a raw state, such as salad and celery, prove extremely inconvenient to the subjects whose case we are now considering. Of course, the quantity of nutriment necessary in any individual case, will be sufficiently indicated by the state of the patient's constitution; but, *generally* speaking, animal is better suited than vegetable food, to such irritable habits, whilst all warm and watery slops prove pernicious, by increasing the susceptibility of the system at large, and by impairing still further the tone of the stomach in particular. Of course, every variety of idiosyncracy will be met with in practice, but, as I have said, *generally* speaking, such patients rather require a generous diet than otherwise, whilst vegetables, especially of the raw and more flatulent sorts for the most part prove highly objectionable. I usually recommend patients to take chocolate or cocoa, or tea, with a large proportion of milk, for breakfast; and if the appetite and digestion will admit of it, a bit of plain cold meat. I prefer plain roast or boiled meat, with a good potato or mashed turnips, for dinner. As to drink, much must depend upon the rank and habits of the individual. Spirits are uniformly objectionable as a habitual indulgence, but a little wine, or wine and water, or a little porter, or ale, may often be allowed with advantage. To conclude, as the stomach is weak, do not load, or oppress it, either by the quantity or quality of the food; as the stomach is irritable, do not excite it by articles that are hot and stimulating; as there is a tendency to flatulency, abstain from vegetables that are wont to produce it; but as there is a general want of tone, and a corresponding morbid susceptibility, let the diet be at once bland and nourishing.

Before I proceed to illustrate our subject by a reference to the patients admitted into the clinical ward, it is right to observe that you are not to expect what can with any propriety be regarded as a complete and permanent CURE of the patients in question. All I pretend, all I profess to do, is to place before you a few examples of which the short duration of my attendance has enabled me to avail myself, and to point out the peculiarities observed in the general character or modification of each, with the variety of treatment founded thereon.

It would indeed be not less unreasonable than vain to expect that the irritable condition of the uterus, and the consequences imparted by its continuance to the general system and to particular parts, should be completely and permanently removed during a short stay in the wards of an hospital. All we can hope to effect in the way of cure, whilst such patients remain with us, is to relieve the more urgent symptoms, and to contribute more or less towards a result which is only to be accomplished by the patients themselves. Accordingly it becomes necessary on discharging them, strongly to impress upon their minds the fact, that their relief is but partial and temporary, and that unless they strictly obey your injunctions, and pay unremitting attention to the precautions you point out, all their ailments and discomforts will certainly return. You ought, therefore, on taking leave, to insist upon the necessity of their using regularly, during the intervals of menstruation, the cold astringent washes already noticed; you ought to urge in strong terms the propriety of their avoiding all violent exertion, and all sources of bodily or mental excitement, whilst the uterus is performing its functions; they should be directed to remain as quiet as possible, and to take some gentle aperient a short time before the expected period; and lastly, they are to be told what sort of diet is most likely to benefit their disorder, and improve their general health.

Such, gentlemen, is the advice you will do well to inculcate in every instance; for, unless you do so, and your advice be attended to, depend upon it any relief you may afford during a professional attendance will be but temporary, whilst each accession of troublesome or painful symptoms will call for medical interference, and tend still further to increase morbid susceptibility, and thereby to aggravate the disorder in the end.

I have said that the complaint assumes different forms in different persons, or even in the same person at different times; that it sometimes presents a general morbid susceptibility both of body and mind such as I have already described, together with more or less disorder of the digestive organs, but without either hysterical paroxysms or local pain, whilst in other instances together with a greater or less degree of the general disorder, we have either a hysterical paroxysm or pain situated in one or more of the parts before specified, or both. Cases of the former description are not often to be found in hospitals; because, in the first place, such patients rarely apply for admission, from a persuasion that their disorder is purely constitutional, and consequently not to be relieved by medicine; and, in the next place, such patients are seldom selected by the attending physician, partly, I fear, from a similar feeling, but chiefly perhaps from a well-founded apprehension, an apprehension founded on past experience, that the remedies usually employed will, in all probability, disappoint both him and his patient.

This general disorder, however, constitutes what may be called the simplest, and in one degree or other by far the most frequent form of the complaint. The extent, too, to which it proceeds, in some instances, without local pain or an actual hysterical paroxysm, is not less surprising to the observer than it is distressing to the patient. I remember a young woman being brought to me by her mother as an out-patient, the state of whose body and mind presented one of the most melancholy pictures of human infirmity I ever beheld. Her mind was so extremely susceptible that she wept from the slightest emotion; in aspect, expression and manner she literally approached to a state of fatuity, whilst her entire frame was so involved in weakness and tremor as almost to resemble an aggravated form of chorea. I was told that her ailments had been coming on and gradually increasing for some years in spite of the various remedies, chiefly of the tonic class, that had been recommended by several practitioners whom she had consulted. I ascertained from her mother that she had all along suffered from profuse leucorrhœa, which, as usual, had never been inquired into, or even hinted at. This young woman made comparatively rapid amendment under the use of the means I have ventured to recommend to your notice.

This is perhaps the extreme of those cases of general disorder unattended by secondary local pain; in practice you will meet with every variety, from the slightest dyspeptic and hypochondriacal tendency to the exquisite form just mentioned.

As presenting a sort of transition from a state of merely general disorder to one in which we have such general disorder associated with an imperfectly developed hysterical paroxysm and local pain, I may direct your attention to our patient, No. 1, in the clinical ward. Her case, as drawn up by our clinical clerk, Mr. Borrett, is as follows:—

"Elizabeth Smith, ætat. 26, has been subject to hysteria, according to her own account, from the age of twenty, when she first became regular, and the affection has gained so great an ascendancy of late as to indispose her completely from her ordinary occupations. She is of a highly nervous and excitable temperament, is flurried and agitated, and goes off into a fit without any obvious cause, which has of late become a very common occurrence. The heart begins to flutter and throb on the slightest emotion of mind, or any bodily effort, or the fit is ushered in by a sudden burst of tears, or by an involuntary fit of laughter, a sense of tightness and oppression is felt at the epigastrium, but there is no ascent of flatus. The fingers of both hands are firmly closed, with numbness of the left side, cramps of the limbs, and general trembling. She is, however, not convulsed, nor does she lose her senses; indeed the fits are very slight, may be prevented by rousing her, and arrested by the cold affusion: they leave, on passing off, a sense of chilliness, which is followed by heat of surface and flushing of the face. She is regular every month, but has more or less leucorrhœal discharge during the intervals, none however at present, bowels costive, pulse 80, small and feeble; she is a dyspeptic subject, troubled with flatulence of stomach, and disturbed by fearful dreams at night."

The most prominent of the permanent symptoms existing at the time of her admission were pain, giddiness, and uneasy sensations within the head, palpitation or pulsation of the heart, pain in the left side, extending occasionally towards the scrobiculus cordis, and remarkable weakness and irritability of the stomach; but with these she presented in a very exquisite degree the general morbid sensibility both of body and mind, to which I have had occasion to direct your attention. The secondary local pain, affecting the left side below the mamma, was by no means considerable, and was moreover very variable in its intensity, it was, however, as is very usually the case, associated with palpitation, or more uniformly with unnatural pulsation of the heart. The hysterical paroxysm was incomplete, it was generally preceded, we were informed, by a tightness and oppression at the epigastrium, but by no distinct globus, whilst neither the mental aberration nor the bodily contortions were so considerable as are commonly observed; but the patient was described as experiencing at the time a numbness of the left side of the body, spasms, impediment in the speech, and other symptoms indicative of disorder of the cerebral functions.

This disturbance in the cerebral functions presents one of the most interesting phenomena of the complaint, and naturally leads to an inquiry into the pathological condition of the brain and nervous system, from whence it proceeds. The symptoms enumerated closely resemble, or indeed are the same as those which attend morbid conditions of the brain of the most serious character; but, as in the complaint under consideration, we know, that, however alarming in appearance, they rarely prove permanent, or rather are rarely followed by serious or fatal consequences, it becomes a most interesting question, both to the pathologist and to the physiologist to determine in what this peculiarity consists; or, in other words, as we have no evidence in many instances of either vascular or organic disturbance, what is the pathological condition of the brain in hysteria?

In making any attempt to solve this interesting question, we are met by difficulties at every step. We still remain in ignorance of the essential properties of cerebral and nervous matter, we are consequently ignorant of the mode in which the functions of the nervous system are performed; whilst we are unfortunately just as ignorant of the ultimate impression made upon that system by remedies given to restore health. In an investigation like the present, therefore, we must be content to take as our only guide general deductions drawn from accumulated facts. From a consideration of these general deductions it would appear, that disorder of the functions of the brain is *chiefly* referrible to three distinct sources :—first, *altered nutrition* of the brain, of its membrane, or of parts adjacent; such as *hardening* or *softening* of the brain alone, or tubercles or tumours of the brain, its membrane, or of the skull. Secondly, *inflammation of the brain or its membrane;* and thirdly, *irregular circulation* within the head, as is observed in threatening or complete simple apoplexy, and in syncope. Such are the three principal sources of direct disorder of the cerebral functions. To which of these sources, then, are we to ascribe the disorder met with in the complaint we are now considering ? Certainly not to organic lesion, the result of altered nutrition : with as little reason can we ascribe it to inflammation, whilst, *in many instances* at least, we have no evidence of its depending upon irregular circulation. What then is the state of the nervous system which characterises hysteria ?

Now, the nervous system is, so far as we know, the only source of sense and motion ; it is through this system that impressions are conveyed to the mind without, and it is through this system alone that the operations of the mind manifest themselves. Providence has so constructed the human machine, and so balanced the susceptibility of the various organs with the impressions made upon them, that in perfect health, and under ordinary circumstances, we acquire such a knowledge of the external world, and experience such emotions, as are best suited to the station in which we are placed. Disease, however, occasionally disturbs or destroys this healthy balance existing between the susceptibility of the sentient organs and the external excitants or stimuli to which they are exposed; so that an immoderate action shall be produced, or an undue, or even painful sensation shall be communicated by ordinary causes—such as in a state of health would have no such effect. This morbid susceptibility of impression, however, is not under such circumstances confined to the sentient organs, but appears to involve the mind in a similar ratio, so that extraordinary emotions are excited by very inadequate causes, or by such as in health would be accounted inadequate.

Analogous to this appears to be the actual state of the nervous system in hysteria, a state, nevertheless, infinitely varied in degree, and consequently infinitely varied in its effects in different individuals. It is a state, however, which, so far as we are able to judge from facts, exists independently of any particular condition of the vascular system, although it is not denied that it may be materially influenced by certain states of that system. It is a morbid state of the nervous system generally, which, by way of distinction, may be said to depend upon some inscrutable change induced by certain causes in the organism or vital endowments of that system, independent of appreciable derangement either of structure or of vascular action—a change, nevertheless, which is sufficiently manifested by the disturbance of the functions of that system, whether that disturbance consist simply of a general morbid susceptibility of body and mind, whether it consist of partial disorder of nervous function in particular parts, or whether it be such as to involve the entire frame in what we designate a hysterical paroxysm. If you ask me why I suppose this state to exist independently of vascular disorder, I can only observe in reply, that I suppose so, because in many instances, even where the disorder of the cerebral functions is most alarming, I have failed to observe, either in the aspect of the patient or in the state of the pulse, any indication of vascular derangement; because, in the next place, I find such disorder of the brain occasionally removed or relieved by remedies which experience teaches us prove injurious in cases where analogous symptoms result from manifest vascular disorder; and, lastly, because, however alarming in appearance, I have seldom known such disturbance prove permanent, or to be followed by serious or fatal results, which cannot be said of cases of a different description. Admitting it then to be *probable* that the morbid condition of the nervous system is of the character I have supposed, you will probably be disposed to inquire *why* I conclude that it proceeds in this case from uterine irritation, and HOW I suppose such uterine irritation to produce it. In reply to the first question, I would say that I suppose it to arise from uterine irritation because of the frequency of their coexistence—because, making due allowance for original differences of constitution, the disturbance of the nervous system is found to be proportionable to the degree or duration of the uterine irritation, and because I find that whatever abates or removes the irritation mitigates or removes the disorder of the brain and nervous system. To the second question, HOW uterine irritation produces this state of the nervous system, my only answer is, I know not.

When Iodine taken in excess produces general morbid irritability, I acknowledge the fact, but cannot explain it; of the *quo modo* I know nothing; when the irritation of a rusty nail, or a thorn, or a burn, involves the entire nervous system in tetanus, I acknowledge the fact, but cannot explain it; when long-continued suckling produces a state of the nervous system analogous to that of which I now treat, I acknowledge the fact, but cannot explain it; so in like manner uterine irritation, by its continuance, disorders the functions of the nervous system, as observed in hysteria, but I cannot explain its *modus operandi*. All these phenomena are so intimately connected with nervous and vital endowments, concerning which we still remain in ignorance, that, were I to attempt more it would only betray me into useless speculation and waste of time.

I have been led into this digression in order, if possible, to convey to your minds such a view of the disorder under discussion as shall prepare you to understand and account for the ever-varying character which it presents according to differences of original constitution, and the susceptibility or degree of derangement of individual organs: a character varying not only in these respects, but moreover through every modification of *disordered action* and *morbid sensation*, from the slightest twitching in the alimentary canal to the most violent convulsions of every voluntary muscle of the body or complete though transitory hemiplegia: from the most trifling uneasiness to the most exquisite torture. But to revert to our case: As she had no severe pain, and as she was troubled with flatulency, I ordered her the Ammoniacal mixture with Magnesia thrice a day, and four grains of Extract of Conium and one of Blue pill night and morning, and to use the Zinc and Alum injection. In this instance, as not unfrequently happens at first, the injection occasioned some smarting, attended with slight leucorrhœal discharge; by persevering in its use, however, these effects soon disappeared. To open her bowels, I ordered her, on the following day, four grains of the Compound Extract of Colocynth, four of the Extract of Rhubarb, and, to obviate smarting, two grains of Extract of Conium. This having failed to act on the bowels, she took, on the following morning, ten grains more of the Compound Extract of Colocynth, which operated, but griped her a good deal. Her stomach was in so weak and irritable a state that the Conium and Blue pill made her sick, and were afterwards withdrawn. Her appetite was almost completely lost; she had long subsisted entirely upon tea and slops, whilst solid food, especially meat, caused a distressing sense of weight and oppression in the epigastrium. Here then we had an exquisite case of general susceptibility, with a state of stomach which not only precluded the possibility of giving those laxatives which prove the least irritating and those medicines most likely to relieve the local pain, but such as prevented the patient taking even a moderate proportion of that food which was most likely to benefit her disorder. Under such circumstances I was compelled to content myself with exhibiting the Ammoniacal mixture, to each dose of which I added a dram of the Tincture of Hops, employing at the same time the injection, and keeping the patient quiet in bed. For the local pain in the left breast we tried at first a Belladonna plaister, and afterwards a blister, but with little relief.

She was admitted on the 31st of December, and on the 10th of January the injection was directed to be desisted from, as the menses were expected, and on the following morning the report was that they had appeared, without being preceded by a hysterical paroxysm as usual, and accompanied by much less pain and suffering than she had previously experienced. All this relief I attributed to the open state of her bowels, and to her being kept quiet in bed; in twelve hours the flow of the menses ceased, when she experienced a considerable aggravation of all her general and local distress. From this state she gradually rallied under the use of Tincture of Hops in Camphor Julep, the use of the injection, and an occasional laxative; so that on the 16th I allowed her a bit of meat, which, however, she herself objected to, in consequence of the uneasiness it was apt to produce in the stomach. It was the extreme irritability of the stomach that compelled me to deviate from the advice I have given you as to the selection of laxatives; here I was obliged to give the Colocynth and Calomel, or the Calomel and Rhubarb in the form of pill, although they always produce much griping.—I did, indeed, order a glyster, but her bowels were so torpid, that I could not trust to it effectually to evacuate them. I need only further add, in reference to this case, that, under the use of the Ammonia and Magnesia, which were resumed, the Extract of Gentian, and the injection, this young woman got gradually better, and left the hospital, I believe, on the 30th, in much better health than she had experienced for a long time. Her general susceptibility was greatly diminished; the disturbance in her head was very much relieved, the hysterical paroxysms ceased to appear, the distension of the stomach and sense of globus were removed; she could digest animal food, and was going on gaining both mental and bodily vigour—but, gentlemen, she was not *cured;* this must be accomplished by her own discretion and by attention to the injunctions given her on quitting our ward.

The next case I shall notice is that of ELIZABETH MARTIN, ætat. eighteen, whose case, as drawn up by Mr. Dashwood, is as follows:[1]

"The catamenia have not been regular for the last two years, occurring at intervals of from six weeks to two months, accompanied with much pain in the back, and lasting a week or ten days; four months ago had pain of the left side, frequent attacks of globus hystericus, sometimes amounting to an absolute hysterical paroxysm; soon after an eruption, having the characters of large blisters filled with serum, arose about the chest, and since which she has had no return of the hysterical symptoms until a week ago, when during the period of menstruation she again became the subject of acute pain of the side, apparently along the course of the arch, and descending portion of the colon; this pain is increased on inspiration, and also by pressure; the bowels are habitually costive; has some cough, accompanied by thin mucous expectoration; pulse 88, soft and quiet; tongue furred in the centre, but clean at the edges; face a little flushed; has frequent palpitations of the heart; has no pain of the head, and has never observed any leucorrhœal discharge."

This young woman appeared to be naturally of a somewhat delicate and susceptible constitution, as indicated by her slender form, her brilliant eye, and thin transparent skin. She had suffered much from painful and excessive menstruation, so that her whole frame was in a remarkably sensitive state; her stomach too was exceedingly weak and irritable, so that no active opiate given to relieve the pain in the track of the descending colon could be retained. Her head too was so much affected with pain and giddiness that even the Conium and Blue Pill could not be persevered in. The cough was purely catarrhal, and had nothing at all to do with the local pain, although the case was presented to our notice as a very good specimen of pleurisy. The pain was indeed, as often happens, in-

---

[1] I ought to apologise to the clinical clerks for transcribing *literally,* cases drawn up without any intention of publication. I have given them thus to avoid bias and partiality.

creased both by a full inspiration and by pressure; but its *situation*, independently of its character, precluded the possibility of its being pleuritic; it varied not only in its intensity, but moreover in its seat, so that, after remaining for some time fixed in the region of the descending colon, it moved a little higher up, and afterwards attacked the sigmoid flexure and bladder, attended with painful micturition.

On the 6th of January, or about ten days after admission, I ordered the injection to be used, and on the following day was told that pain in making water had been experienced for the first time. I was, therefore, disposed to attribute it to the wash, which will, in very susceptible habits, occasionally produce an increase of pain, if used quite cold at first. It turned out, however, that the injection had not been used at all in our absence, so that it had nothing whatever to do with the pain in this situation.

On the 30th the menses appeared, and continued four days without pain, a circumstance which I am disposed to ascribe to her having been kept quiet in bed, to the laxative medicines given her, and in part perhaps to a pill of four grains of extract of poppies, and one grain of ipecacuanha, night and morning, given to relieve the cough. Little attention was paid either to the disturbance in the head, to the palpitation of the heart, or to the local pain, till the cough became relieved under the simplest treatment.

Suffice it to say, in conclusion, that during her stay she was once blooded: that she had a few leeches applied to the region of the sigmoid flexure, a blister to the left side, repeated fomentations, and that on the 12th of February she began the use of the wash, at first tepid. Both the general and local symptoms, however, though variable, proved remarkably stubborn, owing, I believe, to the bad condition into which both the uterus and general system had been brought before we saw her. She remained with us till I quitted the ward, little more than a month, above a fortnight of which we were debarred from the use of proper remedies, by the bronchial affection under which she laboured at the time of admission. We therefore effected less than we otherwise might have done had we had more time.

The next case I shall notice, is one in which we had the pain diffused more generally over the abdomen; the case itself, as well as the outline of the progress and treatment, being furnished me by my friend and able assistant, Mr. Dashwood, as follows:

Jane Russ, æt. 20, admitted January 13, says, that between five and six weeks ago she was admitted into this Hospital, labouring under acute peritonitis, for which she was copiously blooded and leeched, &c., and was discharged about a fortnight from the time of her admission. At the time of her discharge she had pain of the left side still remaining; soon after she was again seized with severe pain over the whole abdomen, with great tenderness on pressure, accompanied by difficulty and pain on micturition. The catamenial discharge appeared nine days since and lasted three days, with acute pain in the back, loins, and down the inside of the thighs. The discharge was in smaller quantity than usual—during the last week for the first time has had frequent attacks of the globus hystericus, with vertigo and headache, but not amounting to an absolute hysterical paroxysm. At present she complains of acute, constant pain, and tenderness in the lower part of the abdomen, which is full and soft; her easiest position is with the thighs flexed. Respiration is slightly hurried, being about twenty-eight in the minute. Pulse 92, and small. Has frequent palpitations of the heart, generally occurring during sleep. Her bowels have lately been much relaxed, but are now regular; tongue slightly furred. The ankles and legs are swollen and painful on pressure; urine is clear and light-coloured, and does not coagulate on the application of heat.—Foveatur Abdom. c. Decoc. Papaveris.—Pil. Conii. c. Hydr. ter die.—Jul. Camph. ter die.

15th.—On falling asleep is suddenly awoke with a sense of suffocation, with *most acute pain under the left breast.* Pulse 60, soft and quiet; bowels open; countenance quite tranquil; abdomen tender on the least pressure; has some pain on micturition. Rep. Fotus et Pil.—Rep. Julep Ammoniæ, ter die.

18th.—The pain of the side and abdomen in general, which had been somewhat less during the last two days, again suddenly became much more intense last night, shooting through to the back, accompanied with nausea. Pulse still remains soft and quiet; tongue a little more furred.

22nd.—The catamenial discharge appeared in a slight degree on the 19th, with much pain in the back, loins, and thighs, but ceased on the same day. The pain and tenderness of the abdomen has varied from being occasionally most severe to perfect freedom from all uneasiness. Complains to-day of great pain during micturition. *Lotio tepida Zinci et Aluminis injicienda.—Rep. Med.*

30th.—The pain and globus hystericus generally recurring during the night, eight drops of Battley's sedative liquor were given at bedtime with the most beneficial effect. The sulphate of zinc was also tried, but was obliged to be withdrawn on account of its exciting nausea. Her appetite now remains good; she sleeps well at night; has had no return of the globus for some days; the pain and tenderness of the abdomen less in severity, and occurring at longer intervals, and during those intervals she is *perfectly free* from all uneasiness.

Although the next case did not occur in the clinical ward, as it was seen by some of you I shall not scruple to quote it as an example of one of the more acute and equivocal forms of general neuralgia affecting the abdomen.

Ellen Jones, æt. 27, admitted into Charity Ward on Wednesday, October 14, 1829.—Is of a sanguine temperament, with red hair. The menses first appeared at the age of twelve, having been preceded by no other inconvenience than drowsiness. They shortly afterwards entirely ceased till the age of twenty, she having had during the whole of the interval a leucorrhœal discharge occurring at irregular periods. She had also during the continuance of the amenorrhœa experienced a slight hemiplegic attack, from which, however, she is now entirely free. When the menses re-appeared at the age of twenty they continued to recur every two or three weeks, often lasting a whole week, exceedingly copious, and mixed with clots, but attended by no pain in the loins, till her marriage a year and nine months ago. Since then she has had neither a child nor a miscarriage, but the menses have ever since been profuse, recurring every two or three weeks, and attended with violent pains in the loins and pain under the left mamma, the latter often going off and returning again. She says she has had medical advice for the last year past, and has been three times blooded—has had leeches applied, and taken medicine, but in spite of which she got gradually worse up to the present time. The pain she now experiences commenced a fortnight ago whilst the menses were flowing, and has gradually increased to its present severity. The pain affects the whole of the abdomen, and is so extremely acute that she cries out from the slightest touch. There is, however, no great expression of distress in the countenance; skin rather warm than hot, with a disposition to moisture; tongue foul and slightly brown; pulse frequent and sharp.

There was a time when I certainly should have set this down as a case of peritonitis, which had come on gradually, and had arrived at its present severity from neglect: an opinion, indeed, which seemed to prevail in the minds of some of those pupils who saw the patient with me. As it occurred, however, in a young female, I was, for the reasons I have mentioned, upon my guard, and expressed my doubts and suspicions to those present. It so happened that whilst we were conversing, the girl actually experienced an imperfect hysterical paroxysm, which appeared in some degree to remove the prevailing scepticism. I ordered twenty leeches to be applied to the belly, and warmth afterwards to encourage the bleeding from the bites, and two grains of calomel and one grain of opium every six hours. As she had had no stool, in the following day I ordered the senna and salts; immediately after the operation of which she had relief, and only felt pain in the track of the colon on each side, which also speedily vanished.

On the 23rd I ordered the cold wash and the ammoniacal mixture.

On the 27th she had a return of the pain under the left mamma, but this soon subsided; and she was discharged on the 31st, free from any pain whatever.

Of course this case is related merely with a view to illustrate the general neuralgic pain of the abdomen, and its connection with the irritable uterus, but without any pretensions to having *cured* the patient. By being kept at rest, and by employing the ordinary means recommended in such cases, she obtained relief, but only, I believe, temporary, unless followed up by attention to the prophylactic measures already pointed out.

The last of our hysterical cases is that of the patient No. 3. Before noticing it, however, it may not be amiss to relate a few examples of those modifications of the complaint which have not been presented to our notice in the clinical ward; and in order to be brief, I will dispense with all comment upon them.

Louisa Burgess, æt. 21, single woman, and apparently of a good constitution, first became regular about the age of sixteen, the appearance having been preceded for five or six months by pain in the back and head, which symptoms were relieved by the change which then took place. She represents herself as having been regular every month, but that the flow has generally been preceded for two days by pain in the loins and legs, giddiness, and occasionally by fits, one of which, she says, lasted three hours. She had four such fits, each preceded by giddiness, and occurring immediately before the menstrual period. Although she has uniformly experienced pain at the period, it is only during the last three months that she has suffered from leucorrhœal discharge. About six weeks ago was seized with violent sick headache, fainting fits, *pain in the right side, under the margin of the ribs,* sometimes extending downwards or over the whole belly, with great tenderness on pressure and occasional palpitation or pulsation of the heart. These pains continued to increase till the time of her admission into the hospital, at which time they were found to be greatly aggravated by pressure, and in some degree by a deep inspiration.

Elizabeth Beauchamp, æt. 24, admitted October 17th, became regular at fifteen, having suffered from pain in the loins for some months previously. Ever since that time she has experienced pains in the loins and thighs, aching in the legs, and coldness of the legs and feet, before and during the discharge. She married about five years ago, soon after which she miscarried in consequence of a fright. She miscarried a second time at the sixth month, but afterwards bore a living child. From the time of her first miscarriage, except while pregnant, she has menstruated every fortnight or three weeks. After the second miscarriage, she first felt *severe pain under the left mamma, with beating or pulsation of the heart,* which has prevailed more or less ever since, but has always been worst during the flow of the menstrual discharge. Five months ago she was suddenly seized with *excruciating pain in the loins,* so severe that she could not stand, but which subsided gradually in three or four hours, leaving a tenderness over the whole of the abdomen. On the 14th of the present month she was seized suddenly with most severe pain in the lower belly. It came on about five o'clock in the morning and lasted three hours, and was attended with a violent desire but with inability to pass her water. After being well fomented, she succeeded in voiding a little, and the pain went off gradually about eight o'clock. On the following day she was seized with intense pains in the *pit of the stomach,* extending across to each side, without vomiting, but accompanied with considerable nausea and pulsation at the scrobiculus cordis. This pain occasionally abated, but never went off entirely, and two days afterwards returned in so severe a degree, that on being requested to see her I immediately admitted her into the hospital. She appeared at the time to be in the greatest agony, she could not bear the slightest pressure on the parts; she leant forward and declared that she must speedily die unless relieved. I gave her ether and laudanum at once with only partial relief. I then had her bled to twelve ounces, ordered calomel and opium, fomentations and glysters, by which on the following day she was free from complaint in the stomach, the pain being then confined to the loins and left side. Her stomach long remained in a very irritable state.

I have within a very short period had three cases of the general neuralgia of the belly, attended with menorrhagia, but such cases present no peculiarity of sufficient interest to warrant me entering into detail. I have, however, observed that instances of this general neuralgia, accompanied by a *tympanitic state of the bowels*, although the most distressing, are happily of least frequent occurrence. I remember a very curious case of this kind, which must be in the recollection of some present. It occurred in the person of a middle-aged woman. I was told that she laboured under chronic peritonitis, and, under the influence of high authority, I concluded that it was so. Her belly was as tense as a drum, exquisitely tender, and presented a most singular appearance from the thousands of leeches which it appeared had at various times been applied. Shortly after I first saw her, however, I was astonished, on visiting her, to find the parietes of the belly quite relaxed and altogether free from pain. This led me to investigate the case narrowly, and I found my patient suffering from irritable uterus and leucorrhœal discharge. Under the appropriate treatment she was relieved, and discharged.

In the next bed to the above, we had a patient supposed to present a case of hepatitis, as indicated by pain under the margin of the ribs of the right side, and a sallow or icteritious aspect of the countenance. This case, however, turned out to be of the same nature as the former, and was treated accordingly.

I have already told you that a case lately occurred in Martha's Ward, which proved so puzzling that I really could not positively determine whether it was peritonitis or mere neuralgia, although from the history of the patient, a young girl of nineteen, who had fallen into prostitution, and from the state of the uterus, there were strong grounds for suspecting it to be merely neuralgia. However, to err on the right side, the case was treated as one of peritonitis. She was blooded to faintness, had calomel and opium, leeches, and so forth. I soon was convinced, however, that I had been wrong. The pain became extremely unsteady, now affecting the ascending, now the descending colon, and then attacking more or less of the abdomen generally, with intervals of perfect ease. She had a very protracted convalescence, which I ascribed to the severe practice into which my mistake had betrayed me. I need only remark further, that I have known patients literally blanched by repeated bleeding and cupping to remove the neuralgic pains described; and on the other hand, I have known them, after months or years of such injurious practice, speedily relieved, and enabled to enjoy the ordinary comforts of life, by a treatment founded on the principles which I have presumed to recommend to your notice.

The clinical case which remains to be noticed is interesting, inasmuch as it is calculated to remove, in some measure, any doubt that might have prevailed in your minds as to the real connection existing between the condition of the uterus and the local neuralgic pains of which I have said so much. It certainly, as far as it goes, tends to confirm the opinion I have advanced, as to these two phenomena standing to each other in the relation of cause and effect. The case is as follows:

"Elizabeth Harris, æt. 26, a woman of irritable habit, subject to frequent pain of the head and left side, with palpitations of the heart on the least exertion—her menstrual periods are regular, and generally attended with but little pain, but the discharge is usually rather profuse, with passage of clots, and lasting a week. Last Tuesday, while the catamenia were flowing, accidentally got wet by upsetting a tub of ice-water over her legs and feet; the discharge ceased almost immediately, but she did not feel any inconvenience until Wednesday afternoon, when she was seized with pain, first commencing at the scrobiculus cordis, but soon became fixed at the lower part of the abdomen. This pain gradually becoming more acute, she came into the hospital on Thursday afternoon (February 11th). At that time it was very severe; the abdomen acutely tender to the touch on first making pressure upon it, but after continuing the pressure it was tolerated. Bowels had not been relieved; pulse 120, but without jerk; tongue slightly furred."

When I first saw her she could not bear even slight pressure on any part of the belly, and as she was rather of a full habit of body, I ordered her to lose sixteen ounces of blood—to take a smart dose of colocynth and calomel, and to have the belly fomented with the chamomile fomentation. As I was satisfied as to the character of the pain, you might perhaps have expected me to order, with moderate depletion, some form of opiate, either by the mouth or by glyster, with a view to relieve the violence of the pain; whilst you were probably surprised at my giving the colocynth and calomel, as I have elsewhere said that such a combination usually irritates, and causes an increase of pain in such subjects. I adopted the practice, however, with a hope that this aloetic compound might, in conjunction with the bleeding and fomentation, bring about a return of the menstrual discharge, which I calculated, with tolerable certainty, would afford speedy relief. In the evening she experienced great aggravation of her pain, evidently from the irritation of the pills, so that at nine o'clock Mr. Dashwood gave her six drachms of castor oil, applied thirty leeches to the belly, and repeated the fomentation. On the following day, her bowels having been freely opened, she obtained considerable relief, the severe pain only coming on at intervals, and being chiefly felt in the region of the sigmoid flexure.

On the evening of the 13th, or the third day from her admission, *she had a slight return of the catamenia, and next morning we found her entirely free from pain.* In the evening, however, *they again ceased*, and at two o'clock in the morning *the pain again returned* in the lower belly. This pain gradually subsided under the treatment adopted, but on the 17th she was represented to be suffering from *acute pain under the margin of the ribs of the right side.* This was supposed by some to be pleuritic, but although it was manifestly increased by inspiration, it varied very much in its intensity, and at intervals was very much aggravated, which with the other symptoms and history of the case, left no doubt in my mind that it was abdominal and purely neuralgic. I applied a blister and gave the conium. On the following morning she was nearly altogether free from pain.

In this case it is scarcely possible not to associate the state of the uterus and the neuralgic pain in the abdomen as cause and effect, in whatever way or by whatever medium the one may bring about the other. This is a relation, indeed, which must be familiar to every one who has had opportunities of witnessing the consequences which so often arise from a sudden suppression of the lochia shortly after delivery, the check being quickly followed by a severe neuralgic pain attacking the abdomen more or less extensively in different individuals; nor do I think it improbable that some of those cases which have been published as anomalous forms of puerperal fever, may have partaken of the same nature, as the painful affections of the abdomen to which I have directed your attention. But this is mere conjecture.

Now, gentlemen, after all that has been said, I dare say you will be disposed to exclaim, "What you have told us is as old as Hippocrates, the old doctrine of hysteria revived!" Be it so; it is to me a matter of total indifference whether my opinions be regarded as novel or as a revival of the old, provided the impression made upon your minds be permitted to accompany you in practice, and tend to secure you credit and advantage. In truth, gentlemen, although I do not believe with Plato that the uterus is a living animal, producing the Protean forms of hysteria by a sort of predatory excursion made into different parts of the body, I nevertheless do agree with Hippocrates and many moderns in ascribing the disease, usually so called, to a certain condition of that organ; but I go further, and only regard what they call hysteria as a mere part and parcel of that extensive series of morbid phenomena which I have endeavoured to point out to your notice in this lecture; for as to the stale objection to the term hysteria, that some of its symptoms, as the globus, occasionally occur in the male, it is scarcely worthy of serious refutation. Since in hysteria the digestive organs amongst the rest are disturbed, it would be singular indeed if some of its symptoms did not present themselves in affections of these organs arising from other causes; and the same may be said of many other apparent anomalies that have been dwelt upon in a similar manner. Neither will I stop to inquire whether any and what sources of mental or bodily irritation are capable of inducing a state resembling that I have described as resulting from the condition of the uterus. My business has been to treat of the latter alone; I leave the rest to others. I shall merely observe that lactation produces a state very analogous to it, if not the same, whilst the two causes combined exert a most powerful and deleterious influence upon the general health, but too extensive to be treated of in this lecture.

Gentlemen, you must extend indulgence to me if I have appeared too egotistic or too sanguine in this matter. If I have been egotistic, you will ascribe it to the influence of my conviction as to the importance of the subject; if too sanguine, you will attribute it to my anxiety to furnish you with those advantages which I know must spring from its attentive and careful investigation. But, gentlemen, if you really require an apology for detaining you so long, I find ample material for that apology in the lively interest which we must all feel in the comfort and happiness of the other sex, doomed as they are, both by the decrees of providence and by human institutions, to drink deep of the bitter cup of suffering. Whatever may be *her* lot in this world, we, as men, must at least acknowledge that whilst Infinite Power gave us being, Infinite Mercy gave us Woman.

CASE

OF

OVARIAN DROPSY

REMOVED BY THE ACCIDENTAL RUPTURE OF THE CYST.

THE subject of the following case was received into the Female Clinical Ward on the 19th of March, 1834. The history, as taken by Mr. Bird, Clinical Clerk at the time, is as follows:

"Ann Binks, aged 44, a tolerably healthy-looking woman, who has been a widow three years, and who has always resided in London, states that her health has usually been very good, till within the last five years. She had one child, twenty-five years ago; has never miscarried; and at the present time menstruates regularly. She has had some cough almost every winter. Five years ago she first noticed a swelling in the left iliac fossa, about the size of an orange, but which rapidly grew larger. For this, and for some general anasarca, she was admitted into this ward, two years ago. She was, at that time, about the size of a seven months' pregnancy. At the end of three months, she was discharged, cured of the anasarca; but still the subject of ovarian dropsy, the abdominal tumour at the period of her discharge equalling the size of a five-months' pregnancy. She, however, was able to go out to service; and continued tolerably well until Monday, the 10th instant; when, being engaged in closing some heavy shutters, and standing upon a pair of steps for that purpose, her foot slipped, and she fell; pulling on her, in the fall, the steps, which struck her across the abdomen. She suffered at that moment excruciating pain; became sick and faint; and was then placed in bed, and had medical assistance procured. She now perceived that the fluid accumulation, which before the accident was circumscribed, had spread over the whole abdomen, rising to the diaphragm, and obstructing respiration. She appears to have then had an attack of peritoneal inflammation; for which she was bled, cupped, and leeched, with some relief; but being unable to perform her duties, she applied to be admitted here this day, the 19th of March. On admission, her face was pale and anxious, surface cold, circulation languid, abdomen distended with fluid; and she complained of great pain, on applying pressure over the abdomen generally, but especially over the lumbar and iliac regions. The bowels were open; tongue very red; she felt thirsty; her pulse was 98, and small; urine copious, and very turbid; and she stated that she had occasionally passed blood by stool since the accident."

It is unnecessary to enter into any minute details respecting either the progress or the treatment of the case. Her symptoms, on admission, were those of general peritonitis, and a moderate degree of bronchitis: for these she was bled to a small amount, fomented, and had calomel, antimony, and opium administered internally. Under this treatment she improved: her gums were reported to be sore on the 22nd; after which the fluid rapidly decreased; so that on the 5th of April there was no fluctuation whatever, and the remains of the ovarian tumour could be distinctly traced, stretching across from one iliac fossa to the other.

On the 7th of April, the descent of the cyst, or some other change induced by the accident, appeared to offer an obstruction to the ascent of the blood through the iliac vein; for she had an attack of phlegmasia dolens, but by no means severe, affecting the left lower extremity. This was subdued, and almost entirely removed in about a fortnight.

This woman is now, December, 1835, servant in a family residing in Cheapside; and although she can still distinguish a small tumour in the left iliac region, she has never experienced any return of the dropsical enlargement.

ON A CERTAIN AFFECTION OF THE SKIN,

VITILIGOIDEA—α. PLANA, β. TUBEROSA.

PREFACE.

UNLIKE the other diseases which Dr. Addison was the first to describe, Vitiligoidea appears to have attracted but little notice from the members of the profession. Its uncouth name, and its great rarity, have alike tended to this result. The view which most medical men seem to entertain concerning it is that it is one of those "anomalous" and exceptional skin-diseases which are occasionally met with in practice, but which are of very little interest to any, save to professed dermatologists, whose systems they seem to set at defiance. This idea, however, is very far from being well founded.

In the paper by Dr. Addison and Dr. Gull, which appeared in the 'Guy's Hospital Reports' for 1850, and to which these remarks are intended as a preface, five cases are recorded. In three of these the symptoms and appearances were most remarkable, and the *nature* of the disease was generally the same in all. Mrs. B— had had jaundice for fourteen months, when patches of a light opaque colour appeared round the eyelids, and on the palms of the hands and on the fingers. Eliza Parachute had had jaundice for fourteen months, when a similar affection showed itself on the hands, and a little later patches were found on the eyelids. Afterwards tubercles appeared on the ears, over the convexities of the finger-joints, and at other parts. Mrs. J— had had jaundice for several years, and for about five years her eyelids had presented a peculiar appearance, similar to that observed in the other two patients.

Since the publication of this paper three cases of the same disease are known to have been under observation at Guy's Hospital. One of these has not hitherto been published. It occurred in a patient of the late Dr. Barlow, admitted into the Clinical Ward in October, 1864. Eight or nine years before, she had had jaundice. From this she is stated to have perfectly recovered; but when she came under observation her face and the upper part of her body were of a dark olive colour, her complexion having originally been fair. For some months she had had dropsy of the abdomen, and to some extent of the legs. There were patches of light buff discoloration in the eyelids of both eyes. These were recognised as being identical with the affection which had been described by Dr. Addison and Dr. Gull.

The second case is that of Laura L—, a patient of Dr. Pavy, who has recorded it in the 'Guy's Hospital Reports' for 1866. She had then had jaundice for three years, and the peculiar changes in the skin had existed for two years. They appeared in both the "plane" and the "tuberose" forms; and affected, in precisely the same way, the same regions as in Dr. Addison's case of Ellen Parachute. As in that case, too, the tubers were not confined to the skin. One, at least, was more deeply seated, being attached to one of the extensor tendons over the knuckles, the skin covering it being quite free and movable. This woman is now attending as an out-patient under the care of Dr. Fagge. She is still jaundiced, and the cutaneous affection has reached a more extreme degree than in any case hitherto recorded. Dr. Fagge has noticed that parts of the mucous membrane of the mouth are altered, in the same way as the skin.

The third case has recently been under Dr. Habershon's care. This patient, a man, æt. 33, has had jaundice for two years, and for about eight months he has observed certain small patches of a creamy yellow colour on the eyelids. These are as yet the only signs of the peculiar change in the skin, but the mucous membrane of the mouth is affected, as in the other patient now under observation. In both these cases the liver is greatly enlarged. In the woman it fills a large part of the abdomen. The enlargement appears to be uniform. The bile-ducts seem to be free, for the motions are of a dark colour. The urine contains bile-pigment.

Our main object, in the preceding observations, is to show how remarkable, and at the same time how constant, are the characters of Vitiligoidea. These cases, taken in conjunction with those recorded in the paper by Dr. Addison and Dr. Gull, surely prove that the disease is no mere curiosity, interesting only to the specialist. On the contrary, it affects various tissues—skin, mucous membrane, tendon, liver. Its pathological nature is as yet a mystery, for it seems to have hardly any tendency to destroy life, and no post-mortem examination of any case has been recorded. Dr. Pavy has shown that the tubercles consist of dense fibrous tissue with interspersed fat-granules, and that the peculiar cream-coloured patches are probably due to the deposition of similar granules in the skin.

It must be admitted, however, that two of the cases (the first and the third) included in the paper of Dr. Addison and Dr. Gull appear not to be examples of the same affection. In these instances there was no jaundice. An examination of the model of the first case, preserved in the museum of Guy's Hospital, confirms the opinion that the disease was of a different nature. Thus, the paper on Vitiligoidea is like Dr. Addison's work on 'Disease of the Supra-renal Capsules,' in containing cases which seem to be essentially distinct from those on which it is mainly based, and which are really examples of an affection previously undescribed. The analogy may aid us in accounting for the erroneous views which have prevailed as to their real character and importance

It may perhaps add to the apparent value of this paper if we remark that we do not know of any recorded example of vitiligoidea, with the exception of those here referred to. Several writers have alluded to the subject in their works upon cutaneous diseases, and some have mentioned in general terms that they have seen cases of a similar kind. But, so far as we are aware, no clinical histories of such cases have been published. It is in most instances tolerably evident, from the way in which the matter is treated, that the writers referred to possessed no well-defined mental images of Vitiligoidea, in the sense in which this term was used by Dr. Addison and Dr. Gull, who invented it.

It is right to state that the preceding paragraphs are in part based upon the remarks made by Dr. Hilton Fagge at a recent meeting of the Pathological Society, at which he exhibited the two cases of Vitiligoidea which are now under observation at Guy's Hospital.

ON A CERTAIN AFFECTION OF THE SKIN,

VITILIGOIDEA—α. PLANA, β. TUBEROSA.

WITH REMARKS AND PLATES.

THE object of this communication is to call attention to a somewhat rare disease of the skin, which, so far as our observations extend, presents itself under two forms : namely, either as tubercles, varying from the size of a pin's head to that of a large pea, isolated or confluent; or, secondly, as yellowish patches of irregular outline, slightly elevated, and with but little hardness. Either of these forms may occur separately, or the two may be combined in the same individual. Under the latter circumstances we are able to trace the connection of the two through an intermediate series of gradations, which clearly demonstrate their essential relations.

It is doubtful whether this disease has been hitherto described. The only account which at all corresponds to it is that given by Willan of vitiligo. He defines vitiligo to consist of " white, shining, smooth *tubercles* arising in the skin, about the ears, neck, and face, terminating without suppuration." Bateman adds, "this disease is somewhat rare, and perhaps but little known." The plate he gives of it is very unlike the appearances presented by the cases we have seen, yet the further description given by him would, to a certain extent, apply to them. "It is characterised," he says, "by the appearance of smooth, white, shining tubercles, which rise on the skin, sometimes in particular parts, as about the ears, neck, and face, and sometimes over nearly the whole body, intermixed with shining papulæ. They vary much in their course and progress : in some cases they reach their full size in the space of a week (attaining the magnitude of a large wart), and then begin to subside, becoming flattened to the level of the cuticle in about ten days. In other instances they advance less rapidly, and the elevation which they acquire is less considerable: in fact, they are less distinctly tubercular. But in these cases they are more permanent; and as they gradually subside to the level of the surface, they creep along in one direction, as, for example, across the face, chequering the whole superficies with a veal-skin appearance. All the hairs drop out where the disease passes, and never sprout again; a smooth, shining surface, as if polished, being left, and the morbid whiteness remaining through life. The eruption never goes on to ulceration." We have extracted the whole description given by this author of vitiligo, that our readers may judge how near it applies to the cases we have to record. As many particulars are named in it which were not present in our cases, and also many are wanting which we have observed, there may be a doubt whether it is here applicable. The two forms of the affection are indicated, and perhaps the want of exact correspondence may be attributable to the want of a sufficiently large number of cases from which to frame a more accurate general description. Believing it to be probable that Willan would have included the cases here recorded under vitiligo, or an allied affection, we have named them accordingly, distinguishing the two varieties by the terms Vitiligoidea tuberosa, and Vitiligoidea plana. We would note here, that authors have not generally used the term vitiligo as Willan and Bateman used it. Alibert limits it to a simple loss of pigment, without alteration of texture, and in this he is generally followed. Neither Alibert nor Rayer gives any description which would apply to the cases we have to record. The "keloide" affection of these authors is altogether of a different character.* It generally exists as a single tumour, arising either spontaneously or upon a cicatrix. Its course is remarkably slow, and leads, subsequently, to contraction and seaming of the skin.

* Keloide " c'est une excroissance faite aux depens du tissu cellulaire de la peau d'une configuration tantôt oblongue et cylindracée, tantôt ovale ou ronde, et bombée d'une couleur rose pale, dure et renitente au toucher, profondement adherente et comme incrustée dans le tegument offrant parfois à sa surface une multitude de petites veines injectées, imitant assez bien la forme d'une cicatrice qui succederait à une forte brulure poussant d'ordinaire vers ses bords des prolongemens bifurqués, qui ont quelque ressemblance avec les pieds d'une tortue ou les pattes d'un crabe, phénomène constant qui justifie complètement la denomination qui a été imposée par M. Alibert à cette tumeur veritablement extraordinaire."—RAYER, *Maladies de la Peau.*

The following is an outline of the history of the cases which we have observed.

Several years ago a young woman, æt. 24, was admitted into the hospital with a peculiar eruption, extending across the nose and slightly affecting both cheeks. It consisted of shining tubercles, varying from the size of the smallest papule to that of ordinary acne. They were of a lightish colour, with here and there superficial capillary veins meandering over them, giving them a faint rose tint. The changes they underwent were very slow; whilst some advanced others subsided. The further course of the case was not ascertained.

The Model 2733^{1st} presents an accurate copy of the appearances.

It was not until the winter of 1848 that our attention was again drawn to the subject, when the following case occurred :—Mrs. B—, æt. 42, of fair complexion and blue eyes, married, mother of eleven children, had been the subject of jaundice for two years, with much pain about the right hypochondria. After the jaundice had lasted fourteen months a change began in the integument, about the eyelids, and in the palms of the hands and flexures of the fingers. The skin was at this time of a lemon tint. The affection of the eyelids consists of *patches of a light opaque colour, with the surface and edges slightly raised*, extending from the middle of the upper lid inwards around the inner canthus, and then outwards along the lower lid to nearly the same extent. There is a small isolated patch at the outer canthus. The disease affects both eyes equally and symmetrically, with the exception of two spots in the right lower lid, about the size of a hemp-seed, more elevated than the rest. The cuticle over the affected parts is healthy. There is no appreciable induration. The patches are more sensitive than the surrounding parts. The capillaries of the cheeks are slightly tortuous. The palms of the hands are of an olive-brown; along the ridges on either side of the flexures, both of the palms and fingers, there is the same opaque, yellowish discoloration. The appearance is much as if the cuticle were thickened, and the disease confined to it; but, on a complete investigation, it is evident that here, as on the face, it is healthy, and that the morbid change is seated in the cutis, which is rather thickened, altered in colour, and has increased sensibility. The disease remained stationary until death, at the end of four years from the beginning of the jaundice. Towards the end the colour of the general surface deepened to a mahogany-brown. No affection of the skin, similar to that described on the face and hands, appeared elsewhere.

The Models 2733^{2d}, and 2733^{3d}, exhibit the appearances presented by this case.

On the 18th August, 1848, a patient was admitted into the hospital, under the care of Dr. Hughes, for diabetes. The following is an outline of his history at the time:—John Sheriff, æt. 27, of middle stature; by occupation a tailor, residing near Kingsbridge, in Devonshire. About six months before he began to pass an unusual quantity of water, feeling at the same time weak and feverish, with a dry, harsh skin. On admission he presented the ordinary symptoms of diabetes; he voided four pints and a half of urine daily, sp. grav. 1050. The treatment pursued was various, but without any obvious improvement. On the 25th January of the following year (1849), the quantity of urine was seven pints and a half, sp. grav. 1042. At this time an eruption somewhat suddenly appeared on the arms, at first apparently of a lichenous character. In the course of ten days it had extended over the arms, legs, and trunk, both anteriorly and posteriorly, also over the face and into the hair; it consisted of *scattered tubercles of various sizes*, some being as large as a small pea, together with shining, colourless papules. They were most numerous on the outside and back of the fore-arm, and especially about the elbows and knees, where they were confluent. Along the inner side of the arms and thighs they were more sparingly present, and entirely absent from the flexures of the larger joints. Besides the compound character produced by the confluence of two or three tubercles, many of the single ones had also a compound character, or appeared to have such, as shown by the prominent whitish nodules upon them. Some looked as if they were beginning to suppurate, and many were not unlike the ordinary Molluscum, but when incised with a lancet they were found to consist of firm tissue, which, on pressure, gave out no fluid save blood. They were of a yellowish colour, mottled with a deepish rose-tint, and with small capillary veins here and there ramifying over them. They were accompanied with a moderate degree of irritation, hence the apices of many were rubbed and inflamed. The nature of the eruption gave rise at the time to much discussion. On its first appearance, some suspected it to have a secondary venereal affection; but there was nothing in the case, nor indeed in the character of the eruption, when carefully examined, to support this view. The only cutaneous affection with which we could associate it, was that of a young woman, whose case we have given above, where the tubercles had occurred in the face only. The eruption continued almost stationary from the end of January to the beginning of March, when many of the tubercles began to subside, having no obvious change in the texture of the skin. At the end of March the patient left the hospital, and the further course of the case was not ascertained. The appearances presented by the eruption in this case are well shown in Model 2733^{6th} in the museum of Guy's Hospital.

Up to this time we had, therefore, these three cases of anomalous affection of the skin, without our being able to do more than suspect a relation between the first and the third. Some further light was thrown upon the subject by the following case.

Eliza Parachute, æt. 33, of middle stature, moderately well nourished; mother of six children; catamenia regular. Her present illness began in 1848; she attributes it to fright, and to a blow received in the left groin whilst attempting to separate two men who were fighting. Two days after this she became jaundiced, and had from time to time severe paroxysmal pains about the hypochondria, lasting for a day or two; the liver being also enlarged and tender. Four months after the commencement of the jaundice (August 4th, 1848) she was admitted into the Hospital under the care of Dr. Hughes. She remained in until the 26th of September, and left much in the same state she was in when admitted. There was at this time nothing complained of beyond the itching and irritation of the skin common in jaundice. The present affection began after the jaundice had continued fourteen months, when she again came under the care of Dr. Hughes. It first appeared in the hands, spreading across the flexures of the joints of the fingers and palms. Soon afterwards a yellowish patch of discoloration began near the inner canthus of the eyelid, and then a precisely symmetrical one at the same part on the opposite eyelid. These patches are very slightly raised, and not obviously indurated; they have extended very slowly. In the early part of the year 1850, two models, 2733^{4th} and 2733^{5th} were made of the case. At this time the patches on the face existed as above described. Along the ridges bounding the flexures in the palm and about the joints of the fingers, there were yellowish, opaque, irregular, and somewhat raised lines. About the thumb, first joints of the fingers, and inner and interior parts of the wrists, there is a gradual transition to a tubercular prominence of the affected parts, and some distinct tubercles exist on the elbow and knee. The diseased parts are tender, so as to give her pain in using a knife to cut bread. The whole surface of the body is of a dull lemon tint. Various means were employed without avail, the disease showing a tendency to progress slowly. Through the kindness of Mr. Startin, under whose care the patient now is, we have been able to observe it up to the present time. The jaundice still remains occasionally deepened by the exacerbation of the hepatic symptoms. The skin is of a dull lemon hue. During the last seven months the affection has become more tubercular, especially about the back of the joints of the fingers of the right hand. The patch of confluent tubercles on the elbow has much increased since the model was taken. Both elbows are similarly affected. There are also tubercles on the right knee, on the superior surface of the great toe, and on both ears. On the hands the gradations from the plane to the tubercular variety are well marked, and the essential relations of the two forms demonstrable. This case has been of the greatest value in enabling us to connect together the cases which had previously occurred. The tubercles about the ears, elbows, joints of the fingers, &c., are of the same character they were in Sheriff's case. They are firm, rather irregular on the surface; have much the appearance, at first sight, of small compound follicles, but on closer inspection are proved to depend upon a change in the cutis. On the surface small venous capillaries may be here and there seen, producing a mottled appearance. In the hands we pass insensibly from the tubercles on the back of the joints to the state described in Mrs. B—'s case, namely, the slightly raised, opaque, yellowish lines about the flexures of the palms and fingers. The further identity of the disease in the two cases is shown by the presence of similar patches about the eyelids in both.

Mrs. J—, æt. 43, of spare frame, and below the middle stature; married; mother of two children, and in good health, until about eight years ago, when her catamenia ceased, probably from fright. After their cessation she was never well, had pains about the right side and through the shoulders; and for several years past, indeed, nearly ever since the commencement of her ailment, has been jaundiced. She was constitutionally of a dark complexion; this has now become a deep olive brown. During the last five years there has been a gradual change in the integument of the eyelids, giving her a strange expression. This affection of the skin began in the upper lid of the left eye, and extended round by the inner canthus to the lower lid. A similar affection then commenced in the right eyelid, and the appearances now presented by the two are remarkably symmetrical. The surface of the affected parts is slightly raised, and the edge defined. The colour is a light opaque yellow, "coloration feuille morte," with a mottling of the faintest rose tint, with a small meandring vessel or two especially on the patches, which are recent and extending. On passing the finger over the surface, there is a slight, yet but very slight, feeling of resistance. The older spots are the most raised. The cuticle is unaffected, and by slight tension of the skin, will be seen to pass unchanged from the normal to the diseased parts. The discoloured patches often smart, and to use the patient's expression, "seem as if gathering;" they have also an increased sensibility.

It will be observed, that, as the disease extended it has run along the lids so as to avoid the Meibomian region, and that in the left eyelid are two sebaceous follicles, enlarged and filled with dark pigment cells; during the last two years a spot of black pigment has appeared on the mucous membrane of the lower lip. The whole course of the disease has been very slow, and its increase, by degrees, almost insensible. There is no affection of the skin of any other part of the body, beyond the change in its colour above indicated. The liver is enlarged, and there is much tenderness about the left hypochondria. The urine contains bile; and the conjunctivæ are of a decided jaundiced tinge.

We have preferred thus recording the cases which have formed the source from which our knowledge of this affection is drawn, to giving any more abstract dissertation upon it; hoping thus that the experience of others may be the more easily compared with our own, for, doubtless, cases of the like kind have occurred to most; although, until attention is especially drawn to a subject, the individual importance of isolated cases is apt to be overlooked. The connection of this affection of the skin with hepatic derangement is obvious, and the exception which occurred in diabetes is of the more interest, inasmuch as modern pathology points to the liver as the faulty organ in this disease. In what way the defective action of the liver operates, can, perhaps, be no further explained at present, than by the general theory of disordered circulating fluids. It is a matter of experience, that various affections of the cutaneous surface, such as numbness, itching, lichen, urticaria, &c., are closely connected with jaundice, depending, probably, upon the direct action of the morbid fluids upon the cutaneous tissues.

The treatment of these cases has been hitherto unsuccessful. They have manifested an inveteracy equal to that of the morbid conditions of the liver, with which they seem to be associated. In the case of Sheriff, many of the tubercles had slowly subsided before he left the hospital, but in the others there was no tendency to disappearance, especially in the patches 'about the eyelids. Some slight benefit seemed to follow the careful and repeated application of Nitrate of Silver, but when the disease is extensive this would hardly be practicable. Mrs. Parachute informs us, that although none of the tubercles have disappeared, yet they are now rather less prominent than they were a year ago.

ON THE

KELOID OF ALIBERT,

AND ON

TRUE KELOID.

PREFACE.

THIS paper on "the Keloid of Alibert" has recently acquired a new interest from the light thrown upon the subject by the researches of Dr. Hilton Fagge ('Guy's Hospital Reports,' 1868). Dr. Fagge, although he considered that Addison was labouring under some error as to the connection of the disease known by this name with that described by Alibert, has arrived at the conclusion that, for this very reason, Addison exhibited, in his description of it, more originality than is generally supposed.

Addison described a very remarkable affection of the skin characterised by an indurated and scar-like condition, and which before that time had found no place in literature. It so often presented an appearance so closely resembling that of a cicatrix from a burn or scald, that the most appropriate term for it which occurred to Addison's mind was "kelis" or "keloid." As, however, Alibert had already adopted the word "cheloide" to represent an affection of the skin which he had described, Addison, with that deference which is the characteristic of the honest, scientific labourer, adopted the term, regarding the condition which he had observed as simply a variety of the form previously described by Alibert. This adoption arose from the belief that the expression "cheloide" by Alibert was derived from the Greek "κηλὶς, a scar or brand," a term so perfectly apposite to the cases which had come under his observation. It is curious that when Addison insisted on spelling the word with a "K" that it should not have occurred to him that Alibert's term, differently spelt, had another derivation.

Dr. Hilton Fagge has now cleared up the difficulty by his discovery of Alibert's original paper, read before one of the societies, in which he does not apply to the disease which he describes the term "κηλὶς" at all.

The paper is a description of cases in which tumours, *having a superficial resemblance to cancer grow on the skin*, and which he at first styled "cancroide;" but recognising the confusion to which this term might give rise, he substituted for it the name "cheloide," derived from "χηλή," a claw of a crab, which was the Greek equivalent.

It would thus appear that Addison's appellation "keloid" was an entirely original expression, adopted under a wrong conception of Alibert's meaning; and that Alibert's "cheloide" and Addison's "keloid" are distinct diseases, the one having a significance differing from that of the other.

Dr. Fagge has, moreover, shown that subsequently to the publication of Addison's paper on this subject, several German and French writers have described remarkable affections of the skin under the names of "scleroma" or "scleriasis," which, in his judgment, are identical with Addison's "kelis."

It will, therefore, be desirable that future writers on the subject should be agreed upon some common designation; and as Addison's title is so like the one already in use for the definition of Alibert's disease, it is possible that the term "scleriasis" may be best adapted to general acceptation.

We do not reprint "The Practice of Medicine" which he published conjointly with the late Dr. Richard Bright, nor the paper on "The Actions of Poisons" by Morgan and Addison, because it is impossible to distinguish what share he had in the publication of either.

ON THE

KELOID OF ALIBERT,

AND ON

TRUE KELOID.

Read before the Royal Medical and Chirurgical Society, Feb. 28th, 1854.

THE term *keloid*, or *kéloïde*, the name given to the singular affections of the integument about to be described, has been variously interpreted; some deriving it from κήλη, a tumour; others, in reference to certain supposed resemblances, from χηλὴ, a crab's claw; or from χέλυς, a tortoise; whilst others, apparently with much greater propriety, derive it from κηλὶς, 'quasi ustione facta macula,' the disease in every instance presenting a greater or less resemblance to some one of the diversified effects left by a burn.

The more immediate object of this very slender communication is to show that the keloid originally described by Alibert, and now so generally recognised, is altogether different in its mode of development, character, and progress, from another disease occurring in the same tissue, and to which, with much greater aptitude, the term keloid may be applied, if we are to regard resemblance to the effects of a burn as its correct interpretation; for I think it will be shown, that whilst the keloid of Alibert and others can hardly be regarded otherwise than as a fibrous tumour developed in the subcutaneous areolar tissue, the other form of disease to which I have alluded, although originating in the same tissue, is of a character and leads to consequences widely different. In order, however, to illustrate and confirm this proposition, it will be necessary to give a description of both diseases; and in so doing, I will, as far as possible, avoid trespassing too much upon the time and attention of the Society.

I propose distinguishing the two diseases in question by the terms "*Keloid of Alibert*," and "*True Keloid*."

KELOID OF ALIBERT.

I have given the name "Keloid of Alibert" to this form of disease, because I believe Alibert to have been the first to discriminate and accurately describe it. In his celebrated work, 'Description des Maladies de la Peau,' will be found a very accurate representation of it, executed with all the artistic skill, and perhaps a little of the exaggeration of colouring, for which that work is so remarkable. He there suggests its holding a middle place between what he so vaguely and indiscriminately calls "dartre" and cancer, and was led in consequence to give it the name of "cancroïde," like cancer; further justifying the appellation, however, by comparing, as others have done, the claw-like rays or processes of the extending disease to the claws of a crab. Since the period of Alibert's original publication, several other writers have furnished cases and commentaries to illustrate the character, progress, or pathology of the disease. Amongst these we find the names of Biett, Velpeau, Cazenave, Coley, and others; but by far the most complete and elaborate essay on the subject has only lately been written by Dr. Dieburg, of Dorpt, and published in the 'Dentsche Klinik' at Berlin, and for a knowledge of which I am indebted to my colleague Mr. Birkett and Dr. Whitley.

The keloid of Alibert first appears in the form of very small, hard, shining, tubercular-looking elevations, of a roundish or oval shape, somewhat firmly set, of a dusky or deep red colour, and generally attended with itching or pricking, shooting or dragging pains in the part. These tumours slowly increase until they attain a height of two or three lines, and comprise an area varying from that of a horse-bean to that of a small almond. So long as they continue to be abruptly prominent, the summit, or even the entire surface of each tumour, instead of remaining uniformly red, not unfrequently presents a pale or blanched appearance, as if from pressure of the increasing tumour upon the cutis situated above it, and which might at first sight be mistaken for some sort of fluid effusion. On close inspection, however, it is found, that from far as this being the case, the tumour displays a hardness, firmness, and elasticity, which almost convey the notion of so much fibro-cartilage, to which indeed it has been not unaptly compared. After an uncertain period, these hard shining tumours become broader, of more irregular outline, and occasionally slightly depressed in the centre. At this time, and sometimes even earlier, by the aid of an ordinary magnifying glass, or by the naked eye, delicate whitish tendinous-looking lines may be perceived, stretching across the surface of the tumours, mingled with minute blood-vessels of a bluish, purplish, or pinkish colour. The extension of each individual tumour now seems to be effected by certain tapering claw-like processes of seldom more than from half a line to a line in breadth, and probably from a quarter of an inch to as much as an inch in length, proceeding from the edges or angles of the expanding tumour. These claw-like processes appear to produce a puckering of the skin; and, as it were, draw the healthy integument into which they pass, towards the original excrescence, and within the influence of the local changes; appearances, nevertheless, which are probably the mere consequences of the stretching and dragging of the integument occasioned by the increasing size of the tumour beneath.

The slow and gradual increase of these tumours may proceed for months or years, and at last attain a size of an inch, an inch and a half, or two inches in length, as much as half an inch or an inch in breadth, and probably an elevation of three or four lines above the level of the surrounding skin. There may be but a single tumour, or there may be several; when more than one, they may be congregated together in the same neighbourhood, or may occupy parts of the integument remote from each other; when of the largest size, the tumour may so stretch and attenuate the integument as actually to protrude beyond it, exposing a red shining excoriated-looking surface. The development of the tumour is occasionally preceded or accompanied by heat, and some degree of puffiness or tumefaction of the surrounding parts, but without redness or other discoloration; a state of things, indeed, which may temporarily supervene at any period of the disorder, either in consequence of some accidental cause of general excitement, some irritation applied to the tumours themselves, or spontaneously, and without any very appreciable cause whatever.

From the very commencement, as has been already observed, the disease is attended with itching and pricking sensations, which, as the former increases, are aggravated to a sense of constriction, or to severe pricking or stabbing pains, which prove extremely distressing to the patient. Under such circumstances, pressing or handling the tumour is loudly complained of; the sufferings of the patient, if a female, are not unfrequently such as to harass her during the whole of the day, and almost completely to deprive her of rest at night.

The morbid deposit which essentially constitutes the keloid of Alibert takes place in the subcutaneous areolar tissue, between the cutis and adipose membrane. The occasional heat and tumefaction of the neighbouring integument, as well as the itching pain and redness of the tumour itself, sufficiently attest that the morbid process is at least accompanied by a degree of vascular excitement nearly allied to inflammation, an inflammatory state which, it would appear, gives rise to a certain amount of adhesion amongst the meshes of areolar tissue around; and, as we know that tumours of considerable size may be developed in the subcutaneous areolar tissue without either uneasiness, pain, or any very obvious change in the appearance of the skin itself, I am inclined to attribute to this accompanying inflammatory and adhesive process the fixed condition of the tumour, the great vascular injection of the superincumbent skin, and the intensity of the local pains, as well as those remarkable puckerings of the integument which attend the increase of the tumour, and constitute the claw-like processes from which some have derived the name "keloid."

The disease most frequently attacks females from the age of 18 to 35 or more, and in a large majority of instances is found situated near the sternum, between or upon the mammæ; it nevertheless occasionally affects the male, and in both sexes has been known to occur on other parts of the body, as the arms, shoulders, neck, belly, or even the head or face. Alibert, as already observed, considered it in some way allied to cancer; an opinion unsupported by any facts with which I am acquainted; whilst others, with perhaps no better evidence, have attributed the predisposition to a scrofulous taint. The development of the disease in different parts of the integument at the same time, or in succession, and its almost certain recurrence after extirpation by the knife or by caustics, clearly point to some peculiar constitutional condition; but what that condition is remains to be ascertained. All that we at present know respecting the exciting cause of the disease, amounting to no more than the fact, that, instead of arising spontaneously, on parts to all appearance previously sound, as is commonly the case, it has not unfrequently been observed to be developed upon and apparently excited by a cicatrix, as of a burn, a boil, or a recent wound, such as that inflicted by the punishment of flogging. To the disease, when occurring under the latter circumstances, Alibert, in a subsequent work, applied the term *spurious* or *false keloid*—the *cicatrix keloid* of Dieburg—a form of the complaint, however, which is sometimes altogether painless.

CASE I. (Pl. 158^{50}, Model 231^{10}, 231^{11})[1] reported by Mr. Pratt. —Susannah Black, æt. 18, a single person, who has been residing with her mother, at Snowsfields, was admitted on the 6th October, 1853, having been transferred from No. 5, Mary, by permission of Dr. Babington, under whose care she had been since the 14th ult.

She is below the middle height; has dark hair, eyes, and complexion; a narrow forehead and heavy expression; but seems intelligent and is highly hysterical, and was formerly apprenticed to a laundress, but not strong enough to continue this occupation.

Her catamenia first appeared at the age of 15, and have recurred regularly since, generally continuing about three days, but with pain in the back and loins, and during the last two years with clots, sometimes of the size of a shilling.

Her father died of diseased heart, but the other members of her family are healthy, and none of her relations have suffered as she now does.

She is marked by the smallpox which she had when three or four years old, but does not look unhealthy, and states that she was always in good health until about two years ago, when, from exposure to cold at Gravesend, while lightly clad, she first became ill, with pain in her head and right side, and at the scrobiculus cordis, shooting thence to the back. Six weeks after this, in Berkshire, having been gradually getting worse in the meanwhile, with loss of appetite and increase of pain, which for a time was so severe as to keep her in a bent position, but occasionally left the scrobiculus cordis and appeared in the loins, she suddenly vomited about a pint of dark clotted blood, after which she became better, but did not lose the pain in her back, and suffered from palpitation of the heart. About three months after this, having returned to London in the interim, the vomiting of blood recurred, and from this time was repeated at intervals, sometimes of two or three months, at

[1] The references are to plates and models in Guy's Hospital Museum.

others of two or three weeks only, until a few days before admission; and once in the hospital, about two weeks since, she brought up a teacupful of blood.

About twelve weeks since she had a gathering in her right breast, which discharged a small quantity of matter; two weeks after, and just as this was healing, her neck, chest, and both breasts swelled a good deal, with a dull aching pain, but without œdema; one week after this, or two or three days after the swelling had subsided, she first noticed two small red pimples on the right breast, at its upper and inner part, which were painful, with a pricking sensation, and tender. Then, about one week after, two other similar raised spots appeared on the left breast, at about the same position, but not symmetrical, and then two smaller ones above these; these all gradually increased in size; but in varying degrees, and, as they did so, at certain stages of their existence became white (?).

There are at present two raised spots on the right breast, nearly oval in shape, and of considerable size; four on the left breast, two large and two small: one on the upper part of the sternum; several at the upper part of the abdomen; and one on the left shoulder; and a cluster of equivocal white spots at the lower part of the back on the right side.

They seem to be in every stage of existence; some small, red, or white; others of varying size, more vascular, generally of a red colour, and marked with small venæ, and traversed by peculiar white lines; but they all change colour occasionally (?) from white to red or even purple, and have a peculiar, firm, and unyielding feel. They have always a dull and aching sensation, converted into a more acute pricking pain by pressure; are more or less raised above the level of the surface, the largest as much as one eighth of an inch, or even more; have irregular margins, much resembling the contraction of a cicatrix, and appear to increase in size by an extension of the white lines which traverse them into the surrounding tissue, like feelers, to which, indeed, their irregular margins are due.

Her chest is well formed, her nutrition good; she seems to be subject to boils; has old cicatrices of venesection on each arm, and a small hard nodule on the left side of the neck, just above the sterno-clavicular joint, resembling an enlarged gland. Her tongue is white and moist: her pulse 80, full and regular; her countenance rather flushed; her bowels, which have been much relaxed, now act about three times daily, the motions being very loose; her appetite is bad; she complains of pain in her head, across the top. The sounds of respiration and of the heart are normal, as well as the resonance of the chest on percussion, but the heart's impulse is strong and heaving, and the pulsations of the aorta felt above the sternum.

CASE II (Model 229, pl. 158[57], pl. 158[54]), furnished by Mr. Whateley, surgeon, of Berkhampstead.—William Garrett, æt. 37, applied to me, about May, 1851, with a small tumour on the skin of the left breast, slightly elevated above the surrounding skin, silvery red in appearance, exquisitely tender, and about one inch in diameter. I recommended its removal, to which he would not then consent. On seeing him about a month afterwards, there was a second appearing, about an inch from the first, and subsequently a third. Such being the case, and fearing that others might still appear, I did not think it advisable to press the operation. He was then sent to Guy's Hospital, at the request of the late Bransby B. Cooper, Esq., in order that a model, &c., might be taken of the tumour in its then state.

After remaining some time, he again came into the country, and was under my care at the West Herts. Infirmary.

The tumour still continuing to grow, and the three having coalesced into one, and having no appearance of any fresh growth in the neighbourhood, I again advised an operation, to which he consented, and I removed it on the 10th of May, 1852, removing with it about a quarter of an inch of the sound skin all round, and fully down to the bone. The wound was dressed with warm water dressing and oil-silk, and was cicatrized. The cicatrix is now sound, and the man in good health.

The tumour, when freshly cut through, in structure, colour, and appearance most nearly resembled a cow's udder.

The slight sketch, No. 4, represents the result of a microscopic examination of the tumour, made, however, under very unfavorable circumstances, by Dr. Habershon, of Guy's Hospital.

A more minute and careful examination of a keloid tumour has been supplied by Dr. Dieburg, of whose account of it the following is a translation:

"On section we observe a dull white colour, a dense tissue in which fibrous structure is visible to the naked eye, and a creaking sound is produced by the knife. On pressure, no fluid exudes in most cases; in a few, a watery fluid is seen, sometimes reddened by blood. This is characteristic, as different from the 'tumores verrucosi cicatricum' of C. Hawkins, from which a peculiar fluid may generally be expressed. Microscopical examination shows the different stages of development of the cells and fibres. We distinguish—1. More or less rounded bodies, the largest 0·05 of a millimètre; in their interior, we see a nucleus, and frequently other molecules. 2. Cells elongated in the direction of one of their diameters, in great numbers: they seem to constitute a characteristic element of all the tumours of 'cicatrix-keloid' (spurious keloid of Alibert). These cells, called by Follin 'elliptical bodies,' are rounded at their extremities, and their sides present central bulging. These cells are about 0·01 millimètres in breadth, and 0·06 in length. They contain a nucleus easily distinguishable by its brightness from the dull surrounding parts. 3. Spindle-shaped bodies, bulging in their centre, and having long, waving appendages. 4. Fibres of cellular tissue and elastic fibres. The fibres of cellular tissue are formed into bundles, which cross each other, and constitute a pretty dense web. The elastic fibres are less numerous and larger than the latter, and are not easily seen without immersion in acetic acid. When a slice of keloid in an early stage of development is placed under the microscope, it is found to consist almost entirely of the spindle-shaped bodies; at a somewhat later period these are seen to have lost their nuclei, and assumed a fibrous appearance: this is most frequent. At a still later period, we see distinct fibrous bundles, crossing each other, and by immersion in acetic acid the elastic fibres become visible. The whole is nourished by a comparatively small number of blood-vessels. The surface is covered by a very thin layer of epidermis, consisting of tesselated cells, very closely pressed together, which require softening before they become visible under the microscope."

The following translation from M. Lebert's 'Traité pratique des Maladies Cancéreuses, et des Affections curables, confondues avec le Cancer,' will probably not be considered out of place.

"Among the cases of spontaneous and multiplied keloid that we have observed, there were two especially curious, in consequence of their multiplicity and extent. In one case, under M. Velpeau, at 'La Charité,' the whole pectoral region of one side was covered with these tumours; many of which were sufficiently large to have reddened and eroded the surface of the skin at their borders.

"In the second case, a child æt. 10½ had a very great number of keloid tumours, developed upon its back, red on their surfaces, and which had formed in the cicatrices which were consecutive to numerous applications of caustic potash, applied to the poor child by a charlatan, who promised to cure, by this method, a scrofulous disease under which the child laboured."

I may add to this passage from Lebert, the fact, that I have myself very recently been consulted in the case of a young lady of about eighteen years of age, upon whose back, shoulders, and breast, I counted as many as thirty keloid tumours. I was told that they originated in the cicatrices of boils which broke out about six or seven months before. From the situation, it had been a case probably of *acne*.

In regard to treatment little can be said. Various internal and external remedies have been tried in vain; and when extirpated by the knife or destroyed by caustics, the disease has, I believe, very generally returned on the seat of the original disease. When, however, the disease has been first developed in a cicatrix—the spurious keloid of Alibert—extirpation has proved more successful, the disease not having again made its appearance in several instances. It has indeed been asserted that the keloid tumour may subside spontaneously, leaving behind a white and depressed cicatrix; but I believe this to be extremely rare, and is in itself a very improbable event, after the tumour has attained any considerable size.

TRUE KELOID.

What I have ventured to call "True Keloid" presents a very remarkable character, and leads to much more serious consequences than the keloid of Alibert. It is a disease, too, which, so far as I know, has not hitherto, with the exception of a slight allusion of Dr. Coley, been either noticed or described by any writer. Like the keloid of Alibert, it has its original seat in the subcutaneous areolar tissue, and is first indicated by a white patch or opacity of the integument, of a roundish or oval shape, and varying in size from that of a silver penny to that of a crown piece, very slightly or not at all elevated above the level of the surrounding skin, and probably unattended, in the beginning, with pain or any other local uneasiness or inconvenience, although a more or less vivid zone of redness surrounding the whole patch, or a certain amount of venous congestion in its immediate vicinity, sufficiently attests the vascular activity or inflammatory process going on in the parts beneath. Occasionally, and especially when the original white patch is of considerable diameter, its surface presents here and there a faint yellowish or brownish tint communicating to the whole spot a somewhat mottled appearance. The slow and insidious change taking place in the areolar tissue either stops and the spot disappears, or it proceeds, and at length begins to declare itself by a feeling of itching, pain, tightness, or constriction in the affected part, and frequently by a certain amount of subcutaneous hardness and rigidity, extending beyond the site of the original superficial patch, although as yet without any necessary change in the appearance of the superincumbent skin. This hardness and rigidity can be distinctly felt, and, especially when situated on the extremities, may sometimes be traced along the course of the neighbouring tendons or fasciæ, or stretching like a cord along the limb, so as to bend or shorten it, and even interfere with natural progression. At length the part originally affected becomes more or less hide-bound, and a similar change taking place around the more superficial fasciæ and tendons, the latter become so tightened, fixed, and rigid, as to be no longer capable of performing their proper functions, and to such an extent, that the whole of a limb, but especially the fingers, may be permanently contracted, bent, and rendered almost as hard and immovable as a piece of wood; thereby impeding progression, distorting the gait, and making the patient a poor miserable cripple for the remainder of his life.

As these changes proceed, the patient continues to experience itching, pain, or a sense of tightness or constriction of the parts, till at length the disease begins to tell upon both cutis and cuticle. The skin, which may have previously presented only a slightly drawn or puckered look, imparting, to a greater or less extent of it, a ray-like appearance, now shrinks or shrivels; it assumes a dry, smooth, or glistening aspect; it undergoes a more decided change of colour, becoming reddish, pinkish, yellowish, or of a dead leaf colour; the cuticle exfoliates; the cutis manifests a tendency to superficial ulceration or excoriation, with consequent scaliness or scabbing, or, when not excoriated, is occasionally surmounted by obscure tubercular or nodular elevations—the whole appearance very closely resembling the remains of an extensive and imperfectly cicatrised burn. From some part of the boundary of the discoloured and shrivelled skin, there may now and then be seen reddish, elevated, claw-like processes, of from half an inch to two inches in length, extending into the sounder integument, and bearing a very exact resemblance to those mentioned as being so characteristic of the keloid of Alibert. It must also be observed that, during the progress of the disease, it is by no means uncommon to find, scattered over various parts of the apparently sound surface, certain oval or roundish and flattened tubercular-looking elevations, which are somewhat hard to the touch, about the size of a split pea or horse-bean, and without any other discoloration than what appears to be the result of accidental friction or irritation.

The above description of true keloid clearly points to some morbid change slowly taking place in the subcutaneous areolar tissue, whilst the itching, pain, and uneasiness experienced by the patient, the red zone surrounding the patch, and the injection of the neighbouring veins, as well as the subsequent appearances presented by the parts affected, would indicate that the morbid process going on in that tissue is one very nearly allied to inflammation, probably of a strumous kind. It would also appear that the inflammatory product, by its subsequent contraction, seriously interferes with the proper nutrition of the cutis, fixes it more or less firmly to the parts beneath, and, when deposited in the immediate neighbourhood of fasciæ and tendons, may, probably, after the lapse of months or years, lead to all those serious inconveniences which I have already described.

I will not abuse the patience of the Society by entering into any speculations respecting the origin and essential nature of this very singular disease; neither is it necessary to dwell upon plans of treatment, further than to observe that, with the exception of iodine, none of the many remedies tried seemed, in extreme cases, to make the slightest impression upon either the appearance or the progress of the disorder. In one instance, however, less advanced, iodine, taken internally, with the simultaneous application of iodine ointment to the affected parts, did appear to arrest the advance of the local changes, and somewhat lessen the rigidity of the affected tendons. Whether the preparations of iodine administered at a very early period of the disorder would prove more effectual, I have had no opportunity of ascertaining, although I am inclined to entertain a strong opinion in its favour.

The following case presents an example of the disease in its earlier stages :

CASE III (Models 222, 223, 224, Pl. 158^{52}), reported by Mr. Towne.—Eliza Watkins, a young woman between 19 and 20 years of age, of ruddy complexion, fleshy and well looking, with light eyes, and hair tending to red, presented herself amongst the out-patients of Guy's Hospital early in June last.

She was in the situation of lady's-maid, and had for some time been residing at Cheltenham. Her general health was good, and at this time apparently undisturbed. She had been suffering from pain and stiffness in the left arm and left leg, for which she was now seeking relief.

The first appearance of the disease had been noticed twelve months previously, when a small white spot, about the size of a shilling, was observed on the left side; but, as neither pain nor inconvenience accrued, no anxiety was felt with reference to it until about eleven weeks prior to her appearance at the hospital, when she first became sensible of pain, attended with a dragging sensation in the left arm and left leg, both limbs being affected simultaneously. Medical assistance was now called in; poppy fomentations were ordered, and for some time persisted in; the disease still making slow but steady progress.

The lady with whom she was living, having occasion to visit London, brought the young woman with her, and took the opportunity of having a second opinion. The case was now treated as a sprain; but the patient, not feeling satisfied, determined to come to the hospital.

The two limbs were in a very similar condition. At this time they presented to the eye but slight indications of the disease, which principally consisted in a hard, drawn, tight look, on the limb being extended; there might, however, be felt, through nearly the whole length of both arm and leg, a rigid band, which gave to the touch the impression of some inelastic substance tightly strained under the integument.

The shoulders presented a mottled appearance, and had several whitish patches interspersed with numerous small tubercular-looking growths. There also existed a chain of spots which nearly surrounded the right nipple, and several others about the neck and breasts. The spot on the left side (described as the first appearance of the disease) had now attained the size of a five-shilling piece, and had thrown out a band upwards towards the cartilage of the ribs, and a second descending towards the pubes.

During the second week in August, I again saw the patient. The pain in the arm and leg had much increased, with "a feeling of shortening" in the limbs affected; and, after sitting for some time, it was with difficulty the foot could now be extended. The band down the arm had become more distinctly expressed, had assumed a slightly tendinous and glistening character, and had thrown out several small lateral processes. A fresh spot had appeared on the upper lid of the left eye, and a second on the outer side of the right leg. Those on the shoulders had become more evident; the larger one had increased in size, become yellowish in colour, glazed on its surface, was hard to the touch; and did not move freely with the surrounding integument.

The next case exemplifies a more advanced stage of the disease:

CASE IV (Model 225, Pl. 158[46]), reported by Mr. King.—Louisa Burston, æt. 11, was admitted, under Dr. Addison, December 8th, 1852.

The patient, who is a very strumous-looking subject, was very strong and healthy as a baby, but was noticed to be slightly rickety when she began to walk; this was between eighteen months and two years of age; but when she was three and a half or four years old she had nothing remarkable about her.

From this time her mother always considered her delicate; but, beyond frequent attacks of ophthalmia, which have deprived her of most of her eyelashes, and appear to have been of a strumous character, she has never suffered any decided illness.

Attention was first directed to the right thigh, about fourteen months ago, on account of complaints on the part of the child of itching in that situation; and this appears to have been so intense, that measures were taken, by tying her hands, &c., to prevent her flaying herself. When first examined, red spots, like flea-bites, were observed thickly studding the inner part of the thigh, about its middle third, but not imparting any feeling of elevation to the finger.

This condition lasted about a fortnight, and was then succeeded by a flaky desquamation of the cuticle, which persisted for two months, during which time the itching continued to be almost intolerable, and when the part was scratched the spots before alluded to would reappear. About or soon after this time the part began to feel thickened, puckered, and hard, and gradually assumed its present appearance.

On the right thigh, about one inch below Poupart's ligament, and nearer the spine of the pubes than the crest of the ilium, commences this singular appearance of the skin, which more nearly resembles the scar left by a burn than anything else. There is a strip, about one inch broad, nodulated and irregular on its surface, and discoloured in a peculiar manner, being partly red, with a predominance of a light brown tint.

This strip of disease proceeds down the thigh, following the course of the sartorius muscle as far as the junction of the upper two thirds with the lower third of the thigh, at which point the most marked discoloration of the skin ceases; but it is found, by examination with the finger, that the same condition of the cellular tissue follows the sartorius to its insertion, and also appears to involve the tendons of the internal hamstring muscles.

In the lower part of the same leg the cellular tissue over the anterior part of the ankle appears to have become involved, and, in particular over the internal malleolus, the integument is firmly attached to the bone.

She has at the present time no peculiar sensation in the affected parts, nor is the use of her leg in walking at all impaired.

Since she has been in the hospital she has taken various medicines, without the slightest perceptible effect.

The next is an instance of the disease in its most aggravated form, reported by Dr. Collingwood. (Models 228, 227, Pl. 158^{55}, 158^{45}.)

Elizabeth Alexander, æt. 12, resides at Ellirfield, in Hampshire, where her father follows the occupation of shepherd. She has a comfortable home, plenty of wholesome food, and attends the village school. The following account is given by the gentleman under whose care she has been for some years:

"When I first saw Elizabeth Alexander she was about 4 years old, and was a robust, healthy child, and has been in good health up to the present time. When nine months old, she, whilst crawling about the house near the fireplace, touched a piece of hot iron with the left arm, between the elbow and wrist, which soon healed up, leaving a slight scar, not so large or deep as that produced by vaccination, and to my own knowledge she has had no other burn or scald. When seven years old she had a mild attack of measles, which was so slight that she was not confined to her bed for a day, and perfectly recovered from it. A few months after the measles, she had a white spot appear on her left side, below the breast, about the size of a fourpenny piece, with a brownish, hard, inelastic state of skin, about the size of a five-shilling piece, surrounding the white spot, and looking as though the skin had been scorched with hot iron, and I asked the question if such had been the case, and was assured by both mother and child that it was not; and in a few weeks I found the brown part of the skin extending to a large circumference, very much more thickened, puckered, and inelastic, giving no pain on pinching up the skin, or on pressure. About six months after, a similar spot made its appearance on the left shoulder, and from a note I made of the case twelve months after, the following were the appearances then presented.

"The shoulder had been affected for a year and a half. About a year and a half ago a white spot appeared upon the shoulder, surrounded by a brownish discoloration, just as though it had been touched with a hot iron, not painful or tender to the touch; it has gradually extended itself around the shoulder-joint and down the upper third of the arm; the skin is shining, hard, and puckered, like the cicatrix from a burn, and the deltoid and other muscles of the shoulder are so diminished as to leave no appearance of their form; the skin thickened, and apparently adhering to the bone, with considerable loss of power and motion, and contraction of the arm.

"About eighteen months after, the hip (left) became affected exactly in the same manner as the side and shoulder. Two years after this, the right shoulder was the seat of mischief of the same nature as that already existing in the other regions."

From the above account, then, it appears that the disease commenced in the left hypochondriac region, next attacked the left shoulder, then the left hip; up to this time, upwards of four years from the first appearance of the disease, the *right* side was unaffected, while nearly the whole of the *left* side was contracted by it. About a year before her admission the right shoulder became the subject of this singular disease, and on a careful examination I discovered upon her right thigh a small patch of puckered skin about as large as a sixpence, the right leg and thigh being otherwise free. Of the existence of this small patch the patient was ignorant, which was suggestive of its being the commencement of the disease in a hitherto sound part; but on careful watching for a period of several weeks, it does not appear that it has increased in size, but rather to have diminished, and the patient affirms that whereas the disease has steadily increased as a whole, individual spots or small patches made their appearance for a short time and have receded again.

On November 10th, 1852, she was admitted into Guy's Hospital, Lydia 18, under Dr. Addison, when she presented the following appearances. The right shoulder is contracted, hard, and tuberculated, the muscles are wasted, and a strip of skin, about one inch and a half wide, extending from the back and shoulder to the inner part of the elbow, is bound to the bone. This part was formerly ulcerated, and the only part which ever was so. It now presents a scaly appearance, and is very hard. The left shoulder is more tuberculated, and more hide-bound, but the disease on this side is more confined to the shoulder proper, and does not extend far down the arm. On the front of each shoulder is a considerable patch, but the chest is otherwise free. Both the elbow-joints are tightly contracted, and permanently bent at nearly a right angle, and the forearms and hands are considerably wasted. The fingers are nearly all bent inwards, and the hands are small, like those of a child six or seven years old.

From the lower angle of the scapula, a semilunar patch (the original disease) runs round to the mesial line, half way between the umbilicus and the nipples. A large irregular patch exists on the left side, immediately below the umbilicus.

The outside of the left thigh is affected throughout its whole length, together with the whole of the left buttock; the left calf is wasted, and measures two inches in circumference less than the right, while the right thigh measures two inches and three quarters more than the left. The left foot is contracted, and the ankle stiff; the toe is pointed downwards, and she walks upon the ball of the toe.

The right thigh is free from the disease, except a small irregular discoloration about as large as a sixpence, on the front of the thigh. These hard shining places have diminished sensibility, and never were painful. None other of the family ever was affected with the same disease. Her general health is excellent.

Case of Keloid disease. Furnished by John Birkett, Esq., Surgeon to Guy's Hospital. (Models 220, 221, Pl. 158[56].)

E. K—, æt. 31, a female, was born in Devonshire, lived some years in the country, but the greater part of her life has been passed in the suburbs of London.

She married at the age of 15 years and 8 months, was confined with her first child at 16 years and 8 months, and never menstruated until after her marriage. She has given birth to eight children, all of whom she suckled with both breasts, although most with the left.

Of regular and temperate habits; she has of late subsisted, since the death of her husband, by working a mangle.

She has always enjoyed good health, with the exception of palpitation of the heart; and her aspect was formerly healthy. At present she is pallid and careworn, from anxiety and a scanty means of subsistence.

I first saw this patient in July, 1851, through the kindness of Dr. Bossey, of Woolwich, who had watched the case.

In December, 1850, and whilst suckling her last infant, she felt an acute pain under the right arm, and observed a curious appearance in the skin of the part.

Now, July, 1851—six months from the discovery of the disease—it occupies a surface of about six inches by three in extent. It is situated on the axillary half of the right mamma, and extends into the right axilla. The skin feels rigid, as if the tissues were of the nature of parchment. It exhibits a peculiar corrugation, resembling that state of the integument known as " cutis anserina," in an exaggerated condition. It is of a peculiar dull, yellowish tint, resembling that of ivory. The part is painful; often there is numbness, and at other times sharp, tingling, shooting pains. The right nipple is retracted—more than usual, for it has never been so well developed as the left.

A patch of the same disease, about one inch square, is developed in the skin of the left axilla.

In the summer of 1852, a third patch was developed, in the skin of the inside of the left arm.

At present—and I saw her in January, 1854—the diseased patches of skin have but little changed their appearances.

They have all increased a little, they all give her more or less pain, and no treatment hitherto adopted has produced any beneficial result.

The patch on the right breast and axilla is longer; the nipple is deeply retracted, indeed invisible, and the gland atrophied. She is much more obese than when I first saw her, and her general health is very good.

The application which seemed to afford her the most relief was the liquor plumbi diacet. dil.

ON THE

DISORDERS OF THE BRAIN

CONNECTED WITH

DISEASED KIDNEYS.

THE object of this communication is threefold :—First, To point out the general character and individual forms of cerebral disorder connected with interrupted function of the kidneys, from whatever cause such interrupted function may arise. Secondly, To show that, in recent as well as in chronic disease of the kidney, the cerebral disorder is not unfrequently the most prominent, and occasionally the only obvious symptom present. And, thirdly, To establish a means of diagnosis, in such obscure or in unsuspected cases, upon the peculiar character of the cerebral affection.

That suppression of urine has the effect of inducing disorder of the brain, has long been familiar to the profession; and was recently illustrated, in a valuable communication made by Sir H. Halford to the Royal College of Physicians. It is also well known to surgeons that mechanical obstruction to the discharge of urine, when long continued, is occasionally followed by a similar result; and Dr. Bright has not failed to demonstrate that there exists, in many instances, a corresponding connection between disorder of the brain and the peculiar change of kidney he has so well described, and so fully illustrated, in his recently published works.

I am not, however, aware that any attempt has hitherto been made to specify with precision, and in detail, the several forms of cerebral disorder arising in connection with disease of the kidney; or that any one has sought to found, upon the character of these cerebral affections, a means of diagnosis available in cases in which, from the absence of the ordinary symptoms of nephritis, of every form of dropsical effusion, and of an albuminous state of the urine, the diseased condition of the kidneys is liable to be altogether overlooked. Experience and observation having led me to the belief that such obscure cases are by no means of very rare occurrence, and that, in the absence of other indications, the renal disease may occasionally be recognised with tolerable certainty by the character of the cerebral disorder alone, I venture to offer what follows as embracing an outline—although a very imperfect outline, I confess—of the general character and individual forms of cerebral disorder arising in connection with interrupted function of the kidneys.

According to my experience, the general character of cerebral affections connected with renal disease is marked by *a pale face, a quiet pulse, a contracted or undilated and obedient pupil, and the absence of paralysis :*—this general character, however, being somewhat modified, in certain cases, by circumstances attending the individual attack.

So far as I have yet been able to observe, the individual forms of cerebral disorder connected with renal disease are the five following :

1. A more or less sudden attack of *quiet stupor;* which may be temporary and repeated; or permanent, ending in death.

2. A sudden attack of a *peculiar modification of coma and stertor;* which may be temporary or end in death.

3. A sudden attack of *convulsions;* which may be temporary or terminate in death.

4. *A combination of the two latter;* consisting of a sudden attack of coma and stertor, accompanied by constant or intermitting convulsions.

5. A state of *dulness of intellect, sluggishness of manner,* and *drowsiness,* often preceded by *giddiness, dimness of sight,* and *pain in the head;* proceeding either to *coma* alone, or to *coma accompanied by convulsions;* the coma presenting the peculiar character already alluded to.

With respect to the first-mentioned form of cerebral disorder connected with renal disease, that of quiet stupor, it is, in its most exquisite form, probably the least frequently met with; the face is pale, the pulse quiet, the pupil natural, or at least obedient to light; and although the patient may lie almost completely motionless, there is no paralysis; for, on attentively watching him for some time, he will be observed slightly to move all the extremities. By agitating him, and speaking loudly, he may sometimes be partially roused for a moment, but quickly relapses into stupor, as before; or it may not be possible to rouse him at all. There is little or no labour of respiration, no stertor, and no convulsions. Slight degrees of it occasionally precede and pass into the next or second form.

This second form of cerebral affection is that of a sudden attack of coma with stertor, or, in other words, apoplexy: it is, nevertheless, different from ordinary apoplexy: it is the serous apoplexy of authors, and presents the usual general characters of cerebral affection depending upon renal disease; for the face, instead of being flushed, is, in almost every instance, remarkably pale; the pulse, though sometimes small, and more rarely full, is remarkably quiet, or almost natural; the pupil, also, although occasionally dilated or contracted, is often remarkably natural in size, and obedient to light; and there is no paralysis. When the labour of respiration is very great, the general character is apt to be modified by an accelerated pulse, and occasionally by a slight flush of the countenance. The coma is for the most part complete, so that the patient cannot be roused to intelligence for a single moment. The stertor is very peculiar, and in a great measure characteristic of this form of cerebral affection connected with renal disease: it has not, by any means, in general, the deep, rough, guttural, or nasal sound of ordinary apoplexy; it is sometimes slightly of this kind; but much more commonly the stertor presents more of a hissing character, as if produced by the air, both in inspiration and in expiration, striking against the hard palate or even against the lips of the patient, rather than against the velum and throat, as in ordinary apoplectic stertor: the act of respiration, too, is usually, from the first, much more hurried than is observed in the coma of ordinary apoplexy. The peculiar stertor coupled with the pale face has, in more instances than one, enabled me to pronounce with confidence the disease to be renal, without asking a single question, and, in cases, too, in which no renal disease whatever had for a moment been suspected.

The third form of cerebral disorder connected with renal disease is that of a sudden attack of convulsions. In this case, also, the countenance is, for the most part, remarkably pale, although, occasionally, slightly flushed at intervals: the pupil is often but little affected: in slight attacks of the kind, the pulse is sometimes singularly quiet; but when the convulsions are severe, and especially when there is such a degree of coma as to be attended with stertor, the heart often sympathises, and the pulse becomes rapid, irregular, and jerking. This form of cerebral affection often passes into the fourth variety; or the cerebral affection shall take on the form of the fourth variety from the commencement: in the latter case, we have merely a combination of the second and third varieties—the coma, hurried breathing, stertor, and convulsions being so blended together, as often to have led to a dispute, whether the affection ought to be designated apoplexy or epilepsy. From what has been already stated, it may in general be very easily recognised as one of the common forms of cerebral disorder connected with renal disease.

The fifth variety is that in which the cerebral disorder makes its approach in a more gradual and insidious manner, usually commencing with dulness of intellect, sluggishness of manner, and drowsiness, gradually proceeding to coma, and more or less stertor, with or without convulsions; these states being at the same time distinguished by the general indications already pointed out. This form of cerebral disorder appears to be that which most commonly supervenes in the progress of the morbid change of kidney described by Dr. Bright; and is very frequently preceded by giddiness, dimness of sight, and pain in the head.

It is scarcely necessary to acknowledge that it remains for future experience and observation to furnish the details of this very imperfect sketch, both as regards the general character and the individual modifications and complications of the cerebral disorders connected with interrupted function of the kidneys. There is, however, one highly interesting question, to which I may briefly advert; and that is, whether there really exists any discoverable relation between the character of the renal affection and of that of the brain—whether the form, permanence, and violence of the cerebral disorder bear any relation whatever to the activity, duration, and extent of the renal disease. This part of the inquiry is, so far as I know, altogether new; and although I am not, at present, in possession of a sufficient number of well-digested facts to justify any very decided or confident conclusions, I nevertheless have imagined that I have already perceived a certain degree of relation between the actual condition of the kidney and the character of the cerebral affection.

Of all the more serious affections of the brain arising in connection with renal disease, the mildest form appears to be that of a tendency to a state of quiet stupor, varying in degree from a mere torpidity of manner and sluggishness of intellect, to complete insensibility to all surrounding objects. Accordingly, I have found this form of cerebral disorder most frequently present in what may be regarded as the least formidable, or more temporary derangements of the kidney. The most exquisite example I ever saw, occurred in a man who at the time presented no dropsical symptom whatever, whose urine was not albuminous, and who made no complaint of pain or uneasiness in his loins. After death, the cortical part of the kidneys was found highly injected, of a deep red or almost chocolate colour, and somewhat softened in its texture; in short, furnishing the strongest indications of a recent nephritic attack in a subdued form: it is also my belief that the same state of things not unfrequently takes place, at an early period, in the progress of scarlatina : we observe an approach to a similar condition of brain in cases of fever, in which the bladder has been allowed to become over-distended : and most assuredly in cases of retention from stricture, and in cases of calculus in the kidney. In all these instances the interruption or impediment to the urinary secretion may be said to be recent or incomplete; and hence, probably, the less degree of severity of the cerebral affection, and the less peril to the patient; for in such instances the symptoms very commonly pass away, and the patient recovers. When, however, the hurtful cause is of an originally nephritic character, the chance of recovery will be less than when the cause of obstruction happens to be merely mechanical and temporary.

The next, in point of severity, of the cerebral affections connected with renal disease appears to be that of convulsions, with comparatively little stertor ;—convulsions, however, which may prove speedily fatal; or which may be repeated an indefinite number of times, but from which the patient very often completely and permanently recovers. Accordingly, I have observed this form of more simple convulsions most frequently associated with what may fairly be regarded as a more exquisite and enduring form of renal disease than that just alluded to : I have observed it most frequently in cases of renal dropsy, subsequent to scarlatina; and in that form of renal dropsy supposed to arise from direct exposure to damp and cold, and commonly known by the name of inflammatory dropsy. As the renal affection has already proceeded to induce dropsy, we cannot but regard it as more fixed and more formidable than in the cases described as being attended with more or less of quiet stupor: and accordingly, instead of merely a certain degree of this latter condition, we have convulsion which may indeed prove fatal, but from which, as already observed, the patient often completely and permanently recovers.

As might have been expected, the most stubborn and intractable, as well as the most fatal cases of cerebral disorder connected with renal disease, are unquestionably those found associated with the chronic and irremediable disorganization of kidney described and illustrated by Dr. Bright. It is nevertheless very far from being true that every such case of renal disease is associated with cerebral disorder: on the contrary, in no very inconsiderable proportion of such cases, even till the period of their fatal termination, no cerebral derangement whatever, or, at least, none of sufficient intensity to attract particular attention, has been observed. Why cerebral symptoms should supervene in one case and not in another, or, in other words, what it is that determines their development in this and in other forms of renal disease, it is impossible in the present state of our knowledge to ascertain: for although a simultaneous diminution of the urinary secretion may occasionally be observed, such a coincidence is by no means constant; the secretion in some instances continuing to flow in a very fair quantity, even at the period of the most formidable attacks of cerebral disorder.

Considering the gravity, permanence, and irremediable nature of the disorganization in this form of renal disease, we might naturally expect that the cerebral disorder, when it does supervene, would, in its constancy, urgency, and intractability, be found in some measure to correspond;—and, accordingly, this has really appeared to me to be the case; the patient suffering repeatedly, or more or less constantly, from heaviness, drowsiness, giddiness, or pain or sense of tightness in the head, and being peculiarly liable to be suddenly seized with the most alarming and most fatal of all the forms of cerebral disorder occurring in connection with renal disease—profound coma and stertor, with or without convulsions.

I have purposely omitted to notice the morbid changes discovered in the brain after death; they are well known to be very often, in appearance at least, extremely slight; and do not, as far as we are yet aware, either in their kind, degree, or situation, offer any explanation of the form or severity of the cerebral disorder which proved the immediate cause of death.

In concluding this brief and imperfect communication, I would again repeat, that my object has been to direct attention to an important and interesting inquiry, rather than to profess a knowledge of the many details and exceptions which must necessarily arise out of future well-directed observation and experience. My pretensions extend not beyond a conviction, that the cerebral affections which occur in connection with renal disease, or other cause of interrupted urinary secretion, present a character generally recognisable by obvious indications; and that these cerebral disorders of renal origin not unfrequently supervene in the absence of ischuria, of dropsy and of those symptoms usually regarded as essential to ordinary nephritis, or, at least, of such a degree of the latter as would be sufficient to attract particular attention, unless supported by the character of the cerebral disorder.

ON THE INFLUENCE

OF

ELECTRICITY,

AS A REMEDY IN CERTAIN CONVULSIVE AND SPASMODIC DISEASES.

It must have occurred to every one engaged in extensive practice, to meet with cases of convulsive and spasmodic disorders affecting females, and with cases of chorea in both sexes, which, whilst they occasioned extreme distress to the patient, and to the patient's friends, have baffled every attempt to afford permanent, and, in some instances, even temporary relief. It was whilst brooding over the humiliating failure of a host of remedies employed in one of such cases, and which will be described in the sequel, that, as a last resource, I determined upon giving electricity a fair trial. I was perhaps, in some measure, induced to do so, in consequence of having an opportunity of securing the assistance of Mr. Golding Bird, in its application. The effect produced by it at once gratified and surprised me, and led to further trials, the results and particulars of which will not, I trust, be deemed altogether unworthy the attention of the profession.

Of course, all claim to originality, or even novelty, is out of the question; electricity having long been enumerated amongst the ordinary remedies applicable to convulsive disorders generally. It is, nevertheless, much to be feared, that many persons, like myself, have been led greatly to underrate its efficacy, either in consequence of its vague and indiscriminate recommendation, or from the inefficient and careless manner in which it has been applied. Certain it is, that, although I have often ordered it myself, and have more frequently witnessed its employment by others, I had never for a moment entertained the belief that it possessed the power over the disorders alluded to, which I am now inclined to concede to it. It is almost superfluous to observe, that the convulsive and spasmodic disorders of females alluded to, are such as, in a large majority of instances, are connected with some irregularity of menstruation: neither is it necessary to dwell upon the difficulty of distinguishing merely functional from organic diseases of the nervous centres; these being matters with which the profession at large is perfectly familiar.

It is but right to state, that the following cases have not been selected because the treatment proved more or less successful, to the exclusion of others in which it failed: on the contrary, they comprise the whole of the cases hitherto subjected to the electrical process about to be described; and hence afford, as far as they go, a fair promise of at least occasional benefit, from the application of this powerful agent in the treatment of the disorders specified.

However undesirable it may be to encumber our Reports with a detailed history of cases, such a mode of procedure becomes almost indispensable in the present instance; but, in order that the length of this communication may not greatly exceed its importance, I shall, without further comment, proceed to lay before the profession a brief description of the mode in which electricity was applied in the subjoined cases, as drawn up by our electrician, Mr. Frederick Bird.

"In the following cases, the form of electricity employed was, with one exception, that elicited by means of the common electrical machine; being made use of either by taking sparks, in the course of the spine; or in the form of shocks, passed through the pelvis.

"In the former case, the patient was seated on an insulated stool, and a metallic connection made between the prime conductor of the machine and the body of the patient: a brass ball, furnished with a wire or chain, in connection with the earth, was then passed upwards and downwards, in the direction of the spine, at a distance of about an inch from the surface. The machine being at this time excited, the patient became charged, and the electricity continued to pass off, accompanied by sparks, to the brass ball, and thence escaping, through the medium of the wire or chain, to the earth: in this manner a rapid succession of sparks could be maintained; and which, in the present instances, was continued until an eruption followed, which assumed very much the appearance of lichen urticatus; the time necessary for its production varying, in different patients, from five to ten minutes.

"For the purpose of passing the shocks, the following method was had recourse to. A large-sized Leyden jar was so placed, that a communication was established between its inside coating and the prime conductor: a "Lane's electrometer" was then fixed into one end of the conductor, so as to admit of the insulated ball of the former instrument being either in contact with, or at any required distance from, the latter: a chain was placed in contact with the outside coating of the jar, and another was attached to the ball of the electrometer; the ends of both of which were furnished with directors, for convenience of application.

"One of the directors was then held upon the symphysis pubis, whilst the other was placed upon the sacrum;[1] by which means the electric current, in performing its circuit, was made to pass through the pelvis. The ball of the electrometer being then placed at a certain distance (generally ⅜ths of an inch) from the prime conductor, motion was given to the machine, and the charging of the jar commenced; and upon a sufficient quantity of electricity being accumulated to enable its discharge to take place by means of the electrometer, the shock was felt. By adopting the use of an electrometer of this kind, the violence of the shocks is made to depend upon the distance of its insulated ball from the conductor of the machine, and not upon the capacity of the jar: hence, it is only necessary to place the ball at a greater or less distance from the conductor, in order to proportion the intensity of the discharge to the nature of the disease or powers of the patient.

"In one case, that of Jessie Wick, the magnetic-electrical machine was made use of, the patient's strength not being sufficient to admit of the ordinary and more powerful form of electricity. The larger helix having been adjusted to the machine, one of the two conducting wires, furnished with a brass disc, was placed over the cervical portion of the spine; whilst the remaining wire, which was a so provided with a disc, was fixed over the lumbar vertebræ: the helix being then slowly revolved, a succession of shocks was ob-

[1] There is a female in attendance for the purpose of adjusting that part of the apparatus more immediately connected with the person of the patient.

tained, which were thus made to traverse the course of the spinal column."

The following cases were taken from the Hospital Books, by my pupils, Messrs. Brereton and Aspland.

CASE I.

JESSIE WICK, aged 17, admitted May 14th, 1837, under Dr. Addison;—a stout, intelligent, well-developed girl, of rather nervous temperament. Her health seems to have been generally very good till the age of 14; at which time, being of remarkably forward sexual development, she began to menstruate. The catamenial discharge immediately became irregular, occurring every fortnight, lasting three days, and accompanied by acute pains in the loins and genital organs. She does not appear, however, to have suffered materially in her health from this irregularity. Two months since, while menstruating, she suffered violent fright; which was immediately succeeded by cessation of the discharge, hysterical fits, and continued trembling of the limbs, much increased by excitement. Bleeding, with blisters and purgatives, was tried; but no relief appearing, she was admitted into Miriam Ward, under Dr. Addison; her condition being as follows :—She is utterly unable to remain in a state of rest for a moment: her limbs, especially the upper extremities, are violently agitated : the mouth is from time to time ludicrously distorted. The most unvarying motion is, a rolling of her clenched hands quickly round each other, with a thrusting forward of the right in a very systematic manner, it occurring after every third revolution. Deglutition and speech but slightly affected. Occasional pain in the head, back, and loins, and under the left mamma: palpitation of the heart, but no abnormal sound. Her spirits are good, but she is fatigued from the continued action of her muscles. The catamenial discharge is expected in two or three days.

The bowels, which had been obstinately costive, having been, by strong purgatives, freed from much highly offensive matter, creosote, in ℥ij doses, was exhibited thrice daily; but violent retching following the seventh dose, it was discontinued. Her spasmodic twitchings, palpitation, and restlessness, were aggravated. Purgatives were continued; and camphor mixture with hydrocyanic acid, and sulphate of zinc, with hyoscyamus and camphor, prescribed. In spite of these remedies, not the slightest abatement occurred in her symptoms: her nights were sleepless, and her irritability excessive. On the 20th of May the catamenial discharge re-appeared, being somewhat after its proper period, scanty, and accompanied by severe bearing-down pains in the back and loins. It ceased again at 2 a.m. of the 21st. From this date, till June 14th, she continued the use of purgatives; with occasional cupping over the loins, blisters along the spine, the above-named mixture, and increasing doses of zinc, reaching at last thirty-six grains daily. Once, on the 23rd of May, the catamenia returned, but in a few hours disappeared again; and at the next and succeeding periods were absent. Not the slightest abatement in her unpleasant symptoms took place: the arms were, perhaps, a little less violently agitated, but the lower extremities, which till late in May had been comparatively quiet, now became much disturbed, and kept up a continued clapping of the feet against the floor: the mouth and eyes, too, were more severely affected than formerly. On the 28th of May, in addition to the other remedies, dashing of cold water on the head, and along the spinal column, was introduced: but this was necessarily omitted for a time, in consequence of the increasing violence of a cough, which had long troubled her, and which was now accompanied by copious bronchial secretion, deeply tinged with blood. She now stated, that some time previous to admission she had been similarly affected. Under depletion and mercurials, this soon yielded; and the former remedies were vigorously applied till June 15th; when the zinc producing constant nausea, the carbonate of iron, in ℨss doses, thrice daily, was substituted for it. On the 23rd of June, the sulphate of iron was substituted in increasing doses, reaching thirty-two grains daily, till August 14th. A very great improvement evidently followed the exhibition of sulphate of iron, assisted by purgatives, chiefly, of calomel and compound extract of colocynth, or aloes and myrrh, and the use of the shower-bath. She was enabled, under this treatment, to sit nearly still in a chair, unless talked to, or otherwise excited; and, with assistance, she could walk pretty well about the ward: there was, however, always a dragging of one or the other of the feet. The squinting, and distortion of the face, had very much subsided, and the peculiar revolving and thrusting forward of the hands diminished: the least excitement, however, would speedily aggravate considerably all her unpleasant symptoms. August 15th, having then been in the hospital exactly three months, she quitted with an intention of going to Ramsgate. For two months from her departure we heard but little of her; but on October 15th she presented herself at Guy's Hospital, among the out-patients, in a much worse condition than she had ever yet been. It is true, there was less agitation of the limbs, but it had only given place to more alarming symptoms. From her friends' statement it appeared, that before reaching her destination she had been seized with epileptic fits, which she had never experienced before. They were represented to have been severe, and, in spite of remedies, had left her in the state in which she appeared: on their approach, the chorea had subsided. She had a foolish imbecile stare, the face dull, she appeared to be almost regardless of surrounding objects, articulation was lost, and she made no attempt, even by signs, to express her feelings. The twitchings of the upper extremities, mouth, and eyes, were less: the inferior extremities appeared paralytic, at least she made no attempt to move them, nor was she able to stand; her bowels were said to be regular; there had been no return of the catamenia. She was immediately put upon ʒj doses of carbonate of iron, and used the shower-bath; and a blister was applied to the spine, along which there seemed to be pain on pressure. The bowels becoming costive, drastic purgatives were again necessary, and croton-oil was given. On the 20th of October the sulphate was substituted for the carbonate of iron; but under this treatment no amendment was produced. After one of her fits, she lay perfectly comatose, scarcely seeming to breathe; and it was not till repeated assafœtida injections had brought away large quantities of fæcal matter, deeply tinged with iron, and stimulating remedies had been used, that her consciousness, and with it articulation, were regained. October 25th, sulphate of zinc in grain doses thrice a day was used, together with a mixture composed of camphor mixture, and ammoniated tincture of valerian; and occasional blisters along the spine. The zinc was rapidly increased, almost ½ gr. daily, till she took 36 grs. per diem, added to which was the Ferri Sulph. But to no purpose: her chorea returned almost as severely as ever; the epileptic fits were frequent, two or three daily, the longest interval between them being three or four days. She could seldom leave her bed; and to guard against her falling from it, side- and foot-boards were necessary. She had now, also, constant headache, which was always aggravated on the approach of a fit; and her spirits became low and desponding. During the epileptic paroxysm, she had violent opisthotonos, the superior part of the occiput almost meeting the heels; firm contraction of the flexors of the fingers and toes, with equally firm contraction of the extensors of the fore-arms and legs; much thick foam from the mouth; stertor; largely-dilated pupil;—the heart's action was quick and tumultuous; with a sound resembling that observed in chlorosis, distinctly heard at its apex. In this condition she would remain from ten to twenty-five minutes; after which, a quick squinting of the eyes, with frequent relaxation and contraction of the flexor muscles of the fingers and toes, would occur: she would then sink into a deep sleep and awake from it after many hours, pale, languid, and perfectly unconscious of all that had happened. It was remarkable, that the slightest touch, when the violence of the paroxysm was subsiding, would instantly reproduce it. In this state she continued many months, with, perhaps, some alleviation of the epileptic attacks, but no improvement in her chorea. The symptoms of the latter were the same as during her former attack—clapping of the feet against the floor, contortions of the mouth, squinting, with revolving and thrusting forward of the hands.

As a last resource, Dr. Addison ordered electricity. Her strength not allowing of the severer application, electro-magnetism was commenced. It caused continued spasm of the flexor-muscles of the arm; so that, till the current was discontinued, she could not relax her grasp of the brass handles. This was commenced on the 20th of April; and on the 10th of the following month she was so far improved, that she could use her needle with tolerable precision: her general health improved; and the fits became slighter, though as frequent as before. Sparks were now drawn from the spine every other day; each exhibition continuing till a vivid eruption was produced. Her improvement was most marked: at the end of a week she was able to walk across the room without assistance: her countenance gradually became less anxious, and the fits declined in frequency.

June 1. Twelve shocks, through the pelvis, every other day, were ordered. The first administration, at the distance of three-eighths of an inch from the conductor, was followed by severe abdominal and pelvic pains, the immediate precursors of the catamenia. The secretion continued for four hours. Shocks to be discontinued.

July 3. Improvement uninterrupted: occasional twitchings are the only indications of chorea. The catamenia have not appeared this month.

A second exhibition of the shocks again occasioned the development of the catamenial function: in six hours it became arrested; after which, she vomited a small quantity of blood.

July 15. She left the hospital entirely free from chorea; though still subject to fits of diminished force and frequency.

CASE 2.—*Chorea.*

EMMA HILLIER, aged 14, stout, plethoric, of dark hair and eyes, admitted June 14th. Her mother states, that, from an early age, she has been subject to epileptic seizures; and four years ago was brought into the hospital with a severe attack of chorea. She was cured at the end of ten weeks; and since then has had periodical returns, the attacks generally observing the recurrence of spring and autumn. The present attack is not severe; but interferes with progression, and slightly with speech. She complains of severe headache; and her temper is irritable.

Elic. Scintillæ Electricæ spinâ dorsi.
Pulv. Rhei c. Cal. gr. xij, p. r. n.

The sparks were drawn off, at the distance of three-eighths of an inch, till the peculiar eruption was produced. After the fourth or fifth trial, the articulation became distinct, and the walk almost quite steady: there were still, however, twitchings in the arms, shoulders, and muscles of the face. At the end of the third week they had all entirely ceased, in the order above enumerated. Two doses of the rhubarb and calomel had been administered during this period.

CASE 3.

WILLIAM SUTTON, aged 14, a healthy looking, but rather small boy, admitted into Lazarus' Ward, under Dr. Cholmley, on May 10th. He states, that he has been thrice affected with chorea; the first attack occurring upon fright, caused by a vicious horse pursuing him. For this attack he was admitted into this hospital, under Dr. Back; and having remained under treatment six weeks, left, cured. Twelve months subsequently he again became the subject of chorea; which, although of longer duration than the former attack, was much less violent. He was then placed under Dr. Bright; and again left the hospital, apparently quite well. In the commencement of the present year, he suffered from a third invasion of the complaint, which he describes as of a more serious character than either of the former. At present, he is unable to remain quiet a moment : he can walk, but his legs bend about frequently while he is doing so. He is continually thrusting his hand along his side. The face is more violently affected than other portions of the body.

Zinc, with purgatives, was administered, but with little benefit, till June 19th; when, upon his coming under Dr. Addison's care, electricity was ordered to be applied daily along the spinal column. This plan was persevered in till July 11th, when, every symptom of chorea having disappeared, and his general health being good, he was presented.

CASE 4.—*Hysterical Paralysis.*

MATILDA SIMMONS, aged 16, of delicate appearance, light brown hair and eyes, mammæ well developed; has been at service in London; and previous to the appearance of the catamenia, a twelvemonth ago, she experienced the ordinary symptoms attending a delay of that function—headache, palpitation, shooting pains, &c.

For six months, the catamenia appeared regularly; cold then occasioned their arrest; and the former symptoms recurred, and were soon accompanied with numbness and coldness of the whole of the left side. This was immediately preceded by a fit; which, from the account of the friends, appears to have been of the truly hysterical character. Ten days before admission, the face, which had hitherto escaped, became numb on the left side, the sight of the left eye became dim, and there was a slight pain in the globe. On the third day there was perfect amaurosis and ptosis. At the time of admission, there was numbness, coldness, and deficient muscular power over the whole of the left side, including the lining membrane of the mouth, nostril, and conjunctiva : in the latter she perceived a burning sensation, but could not appreciate a touch : the pupil was contracted, and not at all obedient to light : she could not raise the upper eyelid. She complained of pain in the head, and giddiness. Bowels torpid.

C. C. Nuchæ.—Emp. Lyttæ postea.
Mist. Magnes. c. Magnes. Sulph. et Tinct. Jalapæ, ʒj t. d.

This treatment was continued for some time, without the slightest benefit.

April 24. Dr. Addison ordered electricity, in the shape of sparks down the spinal column. On the same evening she could bend her fingers: and the 27th recovered, to a certain extent, power over the muscles of the arm.

29th. The third application was made yesterday, with the result of restoring vision, and the power of elevating the upper eyelid, increasing the power of the arm, and improving, to a certain extent, motion and sensation in the leg.

May 4. Improvement continues: she can walk without difficulty; experiences no numbness, but complains of tingling in the fingers of the left hand: the left eye remains as when she was admitted.

June 12. The form of electricity has, during the last ten days, been changed; sparks being drawn from the left eyelid, and shocks passed through the uterus. There is no improvement in the eye, which remains perfectly amaurotic; but the catamenial function was restored yesterday morning, not preceded or accompanied by any particular symptoms. Her general health being good, she has left the hospital at her own desire.

CASE 5.

ANNE BOSHER, aged 21, a stout, short girl, of rather heavy expression, with dark hair and eyes. Her own health, previous to her present illness, had been always good: her family, however, is not a sound one, cerebral diseases afflicting several members of it: a sister has had chorea, but she has no recollection of it. At the age of 16, she lost her mother, which affected her seriously; and about that time a large abscess formed under the left inferior maxilla, which remained some time open, and discharged freely. The catamenia then appeared; and continued regular, in their recurrence and quantity, till the age of 19, when she was suddenly seized with severe pain at the posterior part of the head, accompanied by loss of recollection. A medical man was called in, who used very large depletion. After confinement to bed for six weeks, this attack yielded; but she was left excessively languid, with severe pains in the loins and right leg, irregularity in the catamenial discharge, and diminished muscular power. Some weeks subsequently her right hand became affected with involuntary twitchings, which gradually extended themselves to the whole body, but more particularly to the right side; and at length her agitation was so violent, that straps were required to keep her in bed. Speech and deglutition were greatly impaired, and the muscles of the face and eyelids were in constant action. She had no fit during this period, but the headache was distracting. Medical aid was procured, but to no purpose. She was then admitted into St. Thomas's Hospital; and while there, experienced some sharp epileptic attacks. She remained in that hospital nearly ten months; and then quitted, not much benefited; although at one time she had greatly improved. She was immediately afterwards admitted into Charity Ward, at which time she could with difficulty be retained in bed: she could not walk or, indeed, even stand; neither could she remain quiet for a moment. Her epileptic seizures were frequent and severe: recollection impaired: deglutition and articulation imperfect: the right side is now, and has been always, more seriously affected than the left: headache, with pain in the back and loins. In this state she continued, with very slight improvement, in spite of a host of remedies, comprising cold to the head, blisters and cupping to the spine, drastic purgatives, iron, zinc, and many others.

June 12. Electricity daily, over the spine, was prescribed; the bowels to be regulated by occasional purgatives.

This treatment has been steadily persevered in (being only once or twice interrupted by her fits), and with very evident benefit. The catamenia have re-appeared, although only for a few hours; her headache has considerably decreased, and her memory improved: the fits seldom trouble her, and her muscular power is so much restored, that she can not only walk about the ward without assistance, and without much difficulty, but can carry, without spilling the smallest quantity, a cup pretty full of liquid. There is still, however, a little twitching of the extremities; and the right side remains yet more agitated than the left. The treatment is ordered to be continued.

CASE 6.—*Chorea.*

SARAH KIDD, aged 16, tall, of slight make, dark hair and eyes, swarthy complexion, prominent eyes, and fatuous aspect, admitted Feb. 18th, 1837, into Miriam Ward.

Her family are all suffering from derangement of the nervous centres. One is blind, two epileptic; another, both idiotic and blind. She has always been weakly. The catamenia appeared twelve months ago, preceded by considerable pain; but were arrested in about two hours, by some one in disguise alarming her. They have never appeared since, either naturally or vicariously. Immediately after their cessation, symptoms of chorea came on. The irregular and involuntary motions chiefly affected the neck and face. These increased, so that the strait-jacket was employed, to keep her in bed. Violent muscular agitation, frequent headaches, loss of articulation, and impairment of deglutition, continued for five months, in spite of medical treatment, which chiefly consisted of leeches to the temples and spine, and the use of the shower-bath. Twice during this period she had severe aggravations, lasting three or four hours; during which it was with difficulty that three or four persons kept her in bed. At the end of five months she could walk about, but the twitchings never ceased. The shower-bath was persisted in, till the occurrence of an attack of acute rheumatism.

A fortnight ago, without any assignable cause, the twitchings became more vigorous. At present, the aspect is fatuous, though occasionally wild, and almost maniacal; the symptoms of chorea are well marked and severe; the pupils are dilated; eyesight dim; sweating profuse.

> Pulv. Rhei c. Cal. ℈j, statim.
> Zinci Sulph. gr.j, t. d.
> Pil. Aloes c. Myrrhâ, ℈j, alt. noctibus.

The symptoms remained obstinate. Colchicum in powder was given as a purgative, and continued for some days after the peculiar pea-soup motions had been induced. Valerian and iodide of iron were not more successful.

April 20. Electricity was commenced. Sparks were drawn from the spine, and shocks passed through the pelvis.

29. The sparks induce a variegated eruption, without papulæ, her skin being thick. There is much less twitching of the shoulders, and the hands are more steady. Muscæ volitantes still flat before the eyes. She looks more cheerful, and feels, generally, better.

May 2. Is rapidly improving. She can now walk several steps without falling. Each electric shock produces intense muscular spasm.

Within a day or two from this date, all traces of involuntary muscular action disappeared: her gait, however, remained stiff and ungainly. This was explained by the anatomy of her knees; the patellæ being ill developed, not above a third of their natural size, and seated several inches above their natural position. Very slight flexion is possible; and the attempt causes pain. The dimness of vision and muscæ volitantes have quite disappeared.

20. Presented, quite cured.

CASE 7.—*Chorea treated by Electricity.*

FRANCES SHEAD, aged 12, an active and intelligent girl, of moderate height and stoutness, admitted April 12, 1837, into Miriam Ward.

She has undergone most of the ordinary diseases of childhood: and, without any very apparent cause, has frequently suffered from headache, chiefly confined to the occipital region. During the last three months, this symptom has increased; and she has likewise experienced nocturnal pains in the eyes disturbing her rest, dimness of vision, and muscæ volitantes. The catamenial function is not developed.

On the 24th of March, she was much alarmed by a cat flying at her; and from this period her friends can date slight irregular movements of the hands. These continued for about a week: a paroxysm of pain in the head then occurred, confined to its old spot, and so excessive, that she threw herself down and screamed violently. This ushered in increased muscular agitation, not confined to the arms, but affecting the whole body; progression became difficult; articulation and deglutition much impaired; respiration difficult and laborious; the expiration being attended with a snorting sound.

At present, these symptoms are in full force. The countenance is vacant; there is headache, pain in the right ankle and wrist, general and continued muscular spasm, affecting both sides equally, but most intense in the arms, shoulders, and face. The tongue is broad, slightly furred; its muscles under no control. Pulse 100, feeble: appetite good. A slight "bellows' sound" is audible over the root of the aorta. Upon the neck and back a number of furunculi are forming, apparently owing to stimulating applications which have been used before her admission.

Scammon. gr. v, Hydr. Submur. Sacchar. āā gr.iiss. stat. et alt. auror. Zinc. Sulph. gr. j, Ext. Conii gr. ij, t. d. in forma pil.

April 17. It was necessary to administer the zinc in solution, owing to her inability to swallow anything solid. Articulation is now quite inaudible. The furunculi continue to form; and, when opened, discharge pus, and sloughing cellular membrane. During sleep, which is not much disturbed, there is slight twitching of the fingers. Increased pain and convulsion come on in paroxysms.

20. The zinc has been increased to eight grains thrice a day, without the slightest benefit. Dr. Addison now ordered electric sparks down the spine, every other day.

28. The electricity has been administered four times, and with marked benefit. It is continued each time for about ten minutes, until a vivid eruption appears, closely resembling lichen urticatus, though scarcely so much raised. She can now project the tongue, though only for an instant, and articulate audibly; deglutition is more comfortably performed; she can sit in a chair, and even stand for a short period. The pain in the head is diminished.

May 6. She can now walk without difficulty, and stand on one leg for a short time. The shoulders, arms, and tongue are now the most affected.

12. The electricity is still continued, and the improvement is uninterrupted. She can now walk, without exhibiting any irregular movements. The countenance has entirely lost its fatuous expression. She cannot keep the tongue protruded.

31. Presented quite free from all traces of chorea.

August. Again admitted under Dr. Bright, with a very mild attack of the disease, not sufficient to interfere with her ordinary pursuits. The sulphate of zinc was ordered; and she left, cured, in less than twenty days.

PREFATORY REMARKS

ON

DISEASE OF THE SUPRA-RENAL CAPSULES.

THE brevity of Addison's treatise on "Disease of the Supra-renal Capsules" may excite surprise in the minds of some of our readers unacquainted with its origin and the process of its production. We may, therefore, state that the author contented himself with giving, *more suo*, a brief, lucid, and masterly description of the malady, illustrated by a few cases. It may also interest them to know that, in obedience to the true scientific modesty to which we have before adverted, he had, some years before the publication of his monograph, introduced the subject to a local medical society, and an allusion to the discussion of it which then occurred appears to us of sufficient importance for remarks.

It will be seen from the paper that an attempt to elucidate the nature of a malady which he had styled "*Idiopathic anæmia*," from an inability to associate it with any exact pathological condition, led Addison to the discovery of the diseased state of the supra-renal capsules; he had repeatedly observed an ailment which terminated in fatal exhaustion without any evidence of organic disease; the clinical features of the malady, however, whilst under observation, were prominently bloodlessness, a sense of extreme prostration, and various shades of alteration in the colour of the skin.

It thus necessarily happened that when, in the absence of any other explanatory pathological condition, the supra-renal capsules presented an abnormal state, the association between it and the symptoms exhibited during life took possession of the author's mind, he correlating them as cause and effect; and subsequent writers have spoken of them in the same sense.

We must, however, remind our readers that anæmia and "morbus Addisonii" are not pathologically connected; for whilst in the one case the patient is pale, flabby, breathless, and, perhaps, fat, he is in the other spare, of a brownish hue, manifesting a good colour in his lips and muscles, so that his condition is one rather of asthenia than anæmia.

There is another important point to which the attention of the reader should be attracted; that in transcribing, as we are bound to do, all the pages of Addison's treatise, we necessarily include some paragraphs which the author, had a more extended observation been accorded to him, would have been the first to expunge, under the belief that he was supplying an impulse to the further investigation of the disease. Addison unfortunately inserted some paragraphs implying that if the supra-renal organs were involved in cancer or tubercle, symptoms corresponding to those which he had described would result. All subsequent experience has shown that on this point he was unquestionably in error. He had given an exact and perfect narration of the symptoms attendant upon the disease which now bears his name, but he does not appear to have recognised that the true morbus Addisonii has essential peculiarities of its own; that no other disease or degeneration of the organ is capable of producing the same associated train of symptoms; indeed, that no other primary disease of the organ has been seen. Like most other great discoveries, the full recognition of its importance has been slowly adopted, and the progressive investigation of its clinical nature languidly prosecuted; for nearly a quarter of a century has elapsed since the original paper was read before the South London Medical Society; yet even now it does not find a place in the nosology of some writers, although the evidence of its distinct and essential nature as a malady *sui generis* is conclusive, whether it be deduced from observations made in this country or throughout the continent of Europe.

ON THE

CONSTITUTIONAL AND LOCAL EFFECTS

OF

DISEASE OF THE SUPRA-RENAL CAPSULES.

PREFACE.

IF pathology be to disease what physiology is to health, it appears reasonable to conclude that, in any given structure or organ, the laws of the former will be as fixed and significant as those of the latter, and that the peculiar characters of any structure or organ may be as certainly recognised in the phenomena of disease as in the phenomena of health.

When investigating the pathology of the lungs, I was led by the results of inflammation affecting the lung-tissue to infer, contrary to general belief, that the lining of the air-cells was not identical and continuous with that of the bronchi, and microscopic investigation has since demonstrated in a very striking manner the correctness of that inference,—an inference, be it observed, drawn entirely from the indication furnished by pathology.

Although pathology, therefore, as a branch of medical science, is necessarily founded on physiology, questions may, nevertheless, arise regarding the true character of a structure or organ, to which occasionally the pathologist may be able to return a more satisfactory and decisive reply than the physiologist—these

PREFACE.

two branches of medical knowledge being thus found mutually to advance and illustrate each other.

Indeed, as regards the functions of individual organs, the mutual aids of these two branches of knowledge are probably much more nearly balanced than many may be disposed to admit; for in estimating, we are very apt to forget how large an amount of our present physiological knowledge respecting the functions of these organs has been the immediate result of casual observations made on the effects of disease. Most of the important organs of the body, however, are so amenable to direct observation and experiment, that in respect to them the modern physiologist may fairly lay claim to a large preponderance of importance, not only in establishing the solid foundation, but in raising and greatly strengthening the superstructure of a rational pathology.

There are still, however, certain organs of the body the actual functions and influence of which have hitherto entirely eluded the researches, and bid defiance to the united efforts of both physiologist and pathologist.

Of these, not the least remarkable are the " suprarenal capsules," the *atrabiliary* capsules of Casper Bartholinus; and it is as a first and feeble step towards inquiry into the functions and influence of these organs suggested by pathology, that I now put forth the following pages.

T. A.

24, NEW STREET, SPRING GARDENS;
May 21st, 1855.

ON THE

CONSTITUTIONAL AND LOCAL EFFECTS

OF

DISEASE OF THE SUPRA-RENAL CAPSULES.

It will hardly be disputed that at the present moment the functions of the supra-renal capsules, and the influence they exercise in the general economy, are almost or altogether unknown. The large supply of blood, which they receive from three separate sources; their numerous nerves, derived immediately from the semilunar ganglia and solar plexus; their early development in the fœtus; their unimpaired integrity to the latest period of life; and their peculiar gland-like structure—all point to the performance of some important office: nevertheless, beyond an ill-defined impression, founded on a consideration of their ultimate organization, that, in common with the spleen, thymus, and thyroid body, they in some way or other minister to the elaboration of the blood, I am not aware that any modern authority has ventured to assign to them any special function or influence whatever.

To the physiologist and to the scientific anatomist, therefore, they continue to be objects of deep interest; and doubtless both the physiologist and anatomist will be inclined to welcome and regard with indulgence the smallest contribution calculated to open out any new source of inquiry respecting them. But if the obscurity which at present so entirely conceals from us the uses of these organs justify the feeblest attempt to add to our scanty stock of knowledge, it is not less true, on the other hand, that any one presuming to make such an attempt ought to take care that he do not, by hasty pretensions, or by partial and prejudiced observation, or by an overstatement of facts, incur the just rebuke of those possessing a sounder and more dispassionate judgment than himself.

Under the influence of these considerations I have for a considerable period withheld, and now venture to publish, the few facts bearing upon the subject that have fallen within my own knowledge, believing, as I now do, that these concurring facts, in relation to each other, are not merely casual coincidences, but are such as admit of a fair and logical inference—an inference that, where these concurring facts are observed, we may pronounce with considerable confidence the existence of diseased supra-renal capsules.

As a preface to my subject, it may not be altogether without interest or unprofitable to give a brief narrative of the circumstances and observations by which I have been led to my present convictions.

For a long period I had from time to time met with a very remarkable form of general anæmia, occurring without any discoverable cause whatever—cases in which there had been no previous loss of blood, no exhausting diarrhœa, no chlorosis, no purpura, no renal, splenic, miasmatic, glandular, strumous, or malignant disease.

Accordingly, in speaking of this form in clinical lecture, I perhaps with little propriety applied to it the term "idiopathic," to distinguish it from cases in which there existed more or less evidence of some of the usual causes or concomitants of the anæmic state.

The disease presented in every instance the same general character, pursued a similar course, and, with scarcely a single exception, was followed, after a variable period, by the same fatal result.

It occurs in both sexes generally, but not exclusively, beyond the middle period of life, and, so far as I at present know, chiefly in persons of a somewhat large and bulky frame, and with a strongly-marked tendency to the formation of fat.

It makes its approach in so slow and insidious a manner that the patient can hardly fix a date to his earliest feeling of that languor which is shortly to become so extreme. The countenance gets pale, the whites of the eyes become pearly, the general frame flabby rather than wasted; the pulse, perhaps, large, but remarkably soft and compressible, and occasionally with a slight jerk, especially under the slightest excitement; there is an increasing indisposition to exertion, with an uncomfortable feeling of faintness or breathlessness on attempting it; the heart is readily made to palpitate; the whole surface of the body presents a blanched, smooth, and waxy appearance; the lips, gums, and tongue seem bloodless; the flabbiness of the solids increases; the appetite fails; extreme languor and faintness supervene, breathlessness and palpitations being produced by the most trifling exertion or emotion; some slight œdema is probably perceived about the ankles; the debility becomes extreme. The patient can no longer rise from his bed, the mind occasionally wanders, he falls into a prostrate and half-torpid state, and at length expires. Nevertheless, to the very last, and after a sickness of, perhaps, several months' duration, the bulkiness of the general frame and the obesity often present a most striking contrast to the failure and exhaustion observable in every other respect.

With perhaps a single exception the disease, in my own experience, resisted all remedial efforts, and sooner or later terminated fatally.

On examining the bodies of such patients after death I have failed to discover any organic lesion that could properly or reasonably be assigned as an adequate cause of such serious consequences; nevertheless, from the disease having uniformly occurred in fat people, I was naturally led to entertain a suspicion that some form of fatty degeneration might have a share, at least, in its production; and I may observe that, in the case last examined, the heart had undergone such a change, and that a portion of the semilunar ganglion and solar plexus, on being subjected to microscopic examination, was pronounced by Mr. Quekett to have passed into a corresponding condition.

Whether any or all of these morbid changes are essentially concerned—as I believe they are—in giving rise to this very remarkable disease, future observation will probably decide.

The cases having occurred prior to the publication of Dr. Bennett's interesting essay on "Leucocythæmia," it was not determined by microscopic examination whether there did or did not exist an excess of white corpuscles in the blood of such patients.

It was whilst seeking in vain to throw some additional light upon this form of anæmia that I stumbled upon the curious facts which it is my more immediate object to make known to the profession; and however unimportant or unsatisfactory they may at first sight appear, I cannot but indulge the hope that, by attracting the attention and enlisting the co-operation of the profession at large, they may lead to the subject being properly examined and sifted, and the inquiry so extended as to suggest, at least, some interesting physiological speculation, if not still more important practical indications.

The leading and characteristic features of the morbid state to which I would direct attention are, anæmia, general languor and debility, remarkable feebleness of the heart's action, irritability of the stomach, and a peculiar change of colour in the skin, occurring in connection with a diseased condition of the "supra-renal capsules."

As has been observed in other forms of anæmic disease, this singular disorder usually commences in such a manner that the individual has considerable difficulty in assigning the number of weeks, or even months, that have elapsed since he first experienced indications of failing health and strength; the rapidity, however, with which the morbid change takes place varies in different instances.

In some cases that rapidity is very great, a few weeks proving sufficient to break up the powers of the constitution, or even to destroy life, the result, I believe, being determined by the extent, and by the more or less speedy development of the organic lesion.

The patient, in most of the cases I have seen, has been observed gradually to fall off in general health; he becomes languid and weak, indisposed to either bodily or mental exertion; the appetite is impaired or entirely lost; the whites of the eyes become pearly; the pulse small and feeble, or perhaps somewhat large, but excessively soft and compressible; the body wastes, without, however, presenting the dry and shrivelled skin and extreme emaciation usually attendant on protracted malignant disease; slight pain or uneasiness is from time to time referred to the region of the stomach, and there is occasionally actual vomiting, which in one instance was both urgent and distressing; and it is by no means uncommon for the patient to manifest indications of disturbed cerebral circulation.

Notwithstanding these unequivocal signs of feeble circulation, anæmia and general prostration, neither the most diligent inquiry nor the most careful physical examination tend to throw the slightest gleam of light upon the precise nature of the patient's malady; nor do we succeed in fixing upon any special lesion as the cause of this gradual and extraordinary constitutional change.

We may, indeed, suspect some malignant or strumous disease—we may be led to inquire into the condition of the so-called blood-making organs—but we discover no proof of organic change anywhere—no enlargement of spleen, thyroid, thymus, or lymphatic glands—no evidence of renal disease, of purpura, of previous exhausting diarrhœa, or ague, or any long-continued exposure to miasmatic influences; but with a more or less manifestation of the symptoms already enumerated we discover a most remarkable and, so far as I know, characteristic discoloration taking place in the skin—sufficiently marked, indeed, as generally to have attracted the attention of the patient himself or of the patient's friends.

This discoloration pervades the whole surface of the body, but is commonly most strongly manifested on the face, neck, superior extremities, penis, and scrotum, and in the flexures of the axillæ and around the navel.

It may be said to present a dingy or smoky appearance, or various tints or shades of deep amber or chesnut-brown; and in one instance the skin was so universally and so deeply darkened that but for the features the patient might have been mistaken for a mulatto.

In some cases the discoloration occurs in patches, or perhaps rather certain parts are so much darker than others as to impart to the surface a mottled or somewhat chequered appearance; and in one instance there were, in the midst of this dark mottling, certain insular portions of the integument presenting a blanched or morbidly white appearance, either in consequence of these portions having remained altogether unaffected by the disease, and thereby contrasting strongly with the surrounding skin, or, as I believe, from an actual defect of colouring matter in these parts. Indeed, as will appear in the subsequent cases, this irregular distribution of pigment-cells is by no means limited to the integument, but is occasionally also made manifest on some of the internal structures.

We have seen it in the form of small black spots, beneath the peritoneum of the mesentery and omentum—a form which in one instance presented itself on the skin of the abdomen.

This singular discoloration usually increases with the advance of the disease; the anæmia, languor, failure of appetite, and feebleness of the heart, become aggravated; a darkish streak usually appears on the commissure of the lips: the body wastes, but without the emaciation and dry, harsh condition of the surface, so commonly observed in ordinary malignant diseases; the pulse becomes smaller and weaker; and without any special complaint of pain or uneasiness the patient at length gradually sinks and expires.

In one case, which may be said to have been acute in its development, as well as rapid in its course, and in which both capsules were found universally diseased after death, the mottled or chequered discoloration was very manifest, the anæmic condition strongly marked, and the sickness and vomiting urgent; but the pulse, instead of being small and feeble, as usual, was large, soft, and extremely compressible, and jerking on the slightest exertion or emotion, and the patient speedily died.

My experience, though necessarily limited, leads to the belief that the disease is by no means of very rare occurrence, and that were we better acquainted with its symptoms and progress, we should probably succeed in detecting many cases which in the present state of our knowledge, may be entirely overlooked or misunderstood; and, I think I may with some confidence affirm, that although partial disease of the capsules may give rise to symptoms, and to a condition of the general system extremely equivocal and inconclusive, yet that a more extensive lesion will be found to produce a state which may not only create a suspicion, but be announced with some confidence to arise from the lesion in question. When the lesion is acute and rapid, I believe the anæmic prostration and peculiar condition of the skin will present a corresponding character, and that whether acute or chronic, provided the lesion involve the entire structure of both organs, death will inevitably be the consequence.

If this statement be correct—and I quite believe it to be so—the chief difficulty that remains to be surmounted by further experience in this, I fear, irremediable disease, is a correct and certain diagnosis —how we may at the earliest possible period detect the existence of this form of anæmia, and how it is to be distinguished from other forms anæmic of disorder.

As I have already observed, the great distinctive mark of this form of anæmia is the singular dingy or dark discoloration of the skin; nevertheless, at a very early period of the disorder, and when the capsules are less extensively diseased, the discoloration may, doubtless, be so slight and equivocal as to render the source of the anæmic condition uncertain.

Our doubts, in such cases, will have reference to the sallow anæmic conditions resulting from miasmatic poisoning or malignant visceral disease; but a searching inquiry into the history of the case, and a careful examination of the several parts or organs usually involved in anæmic disease, will furnish a considerable amount of at least negative evidence; and when we fail to discover any of the other well-known sources of that condition, when the attendant symptoms resemble those enumerated as accompanying disease of the capsules, and when to all this is superadded a dark, dingy, or smoky-looking discoloration of the integument, we shall be justified at least in entertaining a strong suspicion in some instances, a suspicion almost amounting to certainty in others.

It must, however, be observed, that every tinge of yellow, or mere sallowness, throws a still greater doubt over the true nature of the case, and that the more decidedly the discoloration partakes of the character described, the stronger ought to be our impression as to the capsular origin of the disorder. The morbid appearances discovered after death will be described with the cases in which they occurred; but I may remark that a recent direction (March, 1855) has shown that even malignant disease may exist in both capsules, without giving rise to any marked discoloration of the skin; but, in the case alluded to, the deposit in each capsule was exceedingly minute, and could not have seriously interfered with the functions of the organs; extensive and fatal malignant disease had, however, affected other parts.

It may be observed in conclusion, that on subjecting the blood of a patient who recently died from a well-marked attack of this singular disease to microscopic examination, a considerable excess of white corpuscles was found to be present.

CASE I.*—Reported by Mr. Thomas Fuller.

James Wootten, æt. 32, admitted into Guy's Hospital, under Dr. Golding Bird, Feb. 6, 1850, has been residing in Long Alley, Moorfields, and is by occupation a baker.

States that he was attacked with cough three years since, which he was unable to get rid of by ordinary remedies, and was finally cured at St. Bartholomew's, after taking pills for one week.

From this time his skin, previously white, began to assume a darker hue, which has been gradually increasing. Twelve months after leaving the above hospital he was laid up from excessive weakness, the result of his cough, which had again appeared, and incapacitated him for his work. He now became an out-patient of

* The cases generally are given in the language and style of their respective reporters.

St. Thomas's, under Dr. Goolden, who cured his cough, and thinking that the colour of his skin depended on jaundice, treated him for that disease, but to no purpose.

He left the hospital in tolerable health, but subsequently lost flesh, and became so excessively weak, the colour of his skin at the same time getting rapidly darker, that he applied for admission here, which was granted him.

Present appearances.—The whole of the skin of the body is now of a dark hue, and he has just the appearance of having descended from coloured parents, which he assured me is not the case, nor have any of his family for generations, that he can answer for, manifested this peculiarity.

The colour of the skin does not at all resemble that produced by the absorption of the nitrate of silver, but has more the appearance of the pigment of the choroid of the eye; it seems to have affected some parts of the body more than others, the scrotum and penis being the darkest, the soles of the feet and palms of the hands the lightest; the cheeks are a little sunken; the nose is pointed; the conjunctiva are of a pearly whiteness; the voice is puny and puerile, the patient speaking with a kind of indescribable whine, and his whole demeanour is childish.

He complains of a sense of soreness in the chest about the scrobiculus cordis.

The chest is well-formed and perfectly resonant; the sounds of the heart are also healthy; there is some slight fulness in the region of the stomach.

The urine is of a proper colour, and he has passed in twelve hours one and a half pints, which has a specific gravity 1008, an acid reaction, and contains neither albumen nor sugar; there is also some pain on pressure in the left lumbar region.

Feb. 8.—Dr. Bird wished a likeness to be taken, so as to be able to watch any alterations in his colour; and considering the case one of anæmia, ordered Lyn. Ferri Iodidi ʒj, ter die; and middle diet.

These he took the whole of the time he was in the hospital, and was discharged in April, rather stronger, but the colour remaining precisely the same.

Shortly after his discharge from the hospital he was seized with acute pericarditis and pulmonic inflammation, under which he speedily sank and died.

The following is a report of the post-mortem examination:—
Lungs universally adherent, the adhesions being very old. The upper lobe of the right lung contained some small defined patches of recent pneumonia, about the size of a crown piece, surrounded by tolerably healthy structure.

The lower lobe was extremely fleshy and without air. The left lung was bound down by old pleuritic adhesions, which were very tough and difficult to be torn through. The substance of this lung was fleshy, and contained but little air.

There was no tubercle or cavity.

The mucous membrane of the bronchial tubes were considerably injected, and, I believe, rather thickened.

The pericardium was distended with a fluid of a deep brown colour, amounting to about half a pint; recent lymph was effused over the whole serous surface.

The liver and spleen were both of weak texture, and easily broken down; the structure of the liver rather coarse.

The gall-ducts pervious.

The gall bladder contained the usual quantity of bile, which was thin, watery, and clear.

The thoracic duct was pervious throughout; and there was no obstruction to any of the veins or arteries that I could discover.

The colour of the blood in the arteries had an unusually dark appearance.

The kidneys were quite healthy and of full size.

The supra-renal capsules were diseased on both sides, the left about the size of a hen's egg, with the head of the pancreas firmly tied down to it by adhesions. Both capsules were as hard as stones. Intestines pale. Lumbar glands natural. No tubercular deposit was discovered in any organ. The head was not examined.

In some of the cases about to be given the capsules merely participated in disease affecting other organs, either of a strumous or malignant character, and it might consequently be doubtful whether the peculiar symptoms depended upon such complications, or upon the special disease of the capsules. In the above instance, however, no such doubt could reasonably be entertained, inasmuch as there was found no abnormal condition whatever of any other organ to which the peculiar symptoms could by any means be attributed. The slow and gradual inroads of the disease, and the remarkable excess of pigment, were sufficiently accounted for by the universality of the change that had taken place in the structure of both capsules; at least, such would be the legitimate conclusion to be drawn from a comparison of the present with other cases about to be related.

CASE 2.—James Jackson, æt. 35. The subject of this case was admitted into the clinical ward, under my own care, November 11th, 1851, and died December 7th, 1851. For the particulars of its history and result I am indebted to my former pupil and present distinguished colleague, Dr. Gull, who was the first to suspect the true nature of the malady during the life of the patient.

A married man, residing at Gravesend, and occupied as a tide-waiter in the customs.

Of a bilious temperament, dark hair, and sallow complexion, which since his illness has much deepened, so that now it is of dark olive-brown. His wife says, " This obvious change in his complexion has been from the beginning of his illness, and gradually came on at that time." There can be no doubt as to this change in the complexion depending upon increase of pigment; for if the lips be turned down the mucous membrane is seen to be mottled by a deposit of pigment, and a closer examination shows that the dark colour of the lips, which at first had the appearance of sordes, is dependent upon the presence of a black pigment, which is not moveable by moistening or washing the lips. There is an expression of anxiety in the face, and the brow is contracted. He gives the following history of himself:

His occupation subjects him to much anxiety; he is exposed to all the vicissitudes of the weather, both night and day; and sometimes his food for weeks together consists of salt provisions.

Eight years ago he had rheumatism, accompanied with great nervous depression; since that time he has enjoyed general good health, with the exception of some attacks of bilious vomiting.

His present illness came on six months ago with headache, vomiting, and constipation. About the sixth day of his illness he became delirious, and was insensible for twenty-four hours. On recovering his consciousness, he was unable to move the fingers of either hand, nor could he move the legs below the knees; the same parts were numb, as was also the tip of the tongue. He continued weak during the whole summer.

Two months ago he resumed his occupation, and remained at it until ten days back, when the old symptoms of headache, vomiting, and constipation returned.

Dr. McWilliam saw him at this time, and found his symptoms to have an intermittent character, and regarded the case as one of miasmatic poisoning, not only from his general symptoms, but also from the dark, poisoned look of his face, not altogether unlike that presented on the approach of the asphyxic stage of cholera. On his admission into the hospital the pulse was extremely small and feeble, the expression of the face pinched, the brows knitted.

He vomited mucus containing altered blood of a dark-brown colour: tongue clean; epigastric region full, especially towards the left side, where he had some twitching pain and slight tenderness on pressure; urine natural in colour and quantity, of a light brown colour, not coagulable by heat. He went on, day by day, with but slight symptoms of change. Skin cool; pulse moderate in frequency, but extremely feeble, so as scarcely to be felt at the wrist. On several occasions the depression was so great as to require the exhibition of decided stimulants. There was a continued tendency to sickness. The abdomen soft, with marked aortic pulsation. Bowels constipated; chest everywhere resonant; heart's sound normal; extent of dulness on percussion not increased. Slight traces of intermittence in the symptoms; the surface in the evening being cool, or even cold, and the following morning warm, as if from reaction.

Probable diagnosis.—The epigastric tenderness and pulsation, with frequent vomiting, and the ejected mucus and altered blood, point to an inflammatory condition of the gastric mucous membrane. But what condition of system is it which favours the production of black pigment? Is it some affection of the liver; or is it, as Dr. Addison supposes, disease of the supra-renal capsules?

Sectio-cadaveris.—The lining membrane of the stomach was finely injected into minute puncta and stellæ of a bright red colour, with two or three spots of ecchymosis. The structure of the membrane was thickened and pulpy, and the surface covered with tenacious mucus. In some parts there were irregular superficial abrasions; these appearances of the mucous membrane becoming very distinct by examining it under water by the aid of sunlight, and seeming, moreover, unequivocally to demonstrate the existence of a gastritis. The brain, lungs, heart, spleen, liver, and kidneys were normal.

The supra-renal capsules contained, both of them, compact fibrinous concretions, seated in the structure of the organ; superficially examined, they were not unlike some forms of strumous tubercle. The slow and insidious approach and progress of the constitutional loss of strength, the extreme feebleness of the pulse, the absence of all evidence of any lesion sufficient to account for the patient's declining condition, the loss of appetite, the uneasiness and irritability of the stomach, and the indications of disturbed cerebral circulation, were all so strongly marked, and so exactly corresponded in kind with what have been observed to accompany the most extensive disease of the capsules, that, coupled with the excess of dark pigment in the integument, we did not hesitate to anticipate, with much confidence, an extensively diseased condition of these organs.

CASE 3.—Reported by Mr. Williams.

Henry Patten æt. 26, a carpenter and window blind maker, residing at 13, Brandon Street, Walworth, was admitted Nov. 9, 1854, having been for some time an out-patient under Dr. Rees.

His habits have been somewhat intemperate; his drink chiefly malt liquor and spirits.

With the exception of a sister, who died of phthisis, all his relations are healthy. He has been married four years. The patient states that up to six months ago he enjoyed very good health, but then began to be troubled with what he calls "rheumatic" pain in the right leg, which, without laying him up, gradually extended to his hips and side, and thence to the bottom of the spine. His back latterly has been very tender, a jerk or jarring movement giving him great pain at that part. He has noticed his hips to have become dark-coloured for the last three months, and more lately his face to be similarly discoloured in patches. For the last month he has discontinued work on account of attacks of giddiness and dimness of sight, accompanied by a peculiar pain at the back of the head and partial loss of consciousness. These attacks would occur several times in the course of the day, upon any unusual exertion, always while in the standing posture, and were instantly relieved by sitting or lying down. Since he has discontinued his employment they have only occurred on getting out of bed in the morning. It is for the pains and tenderness at the back, and occasional attacks, as above described, with general debility, that he has been attending this hospital as an out-patient.

Present condition.—The patient presents a highly strumous appearance, being thin, pale, and the hair dark and dry. Over the face and forehead, which are of a general yellowish hue, are several patches of darkened skin, and similar black patches on the lips. There is angular curvature at the second, and great tenderness on pressure over the upper three lumbar vertebræ; he complains also of pain at this part upon moving in bed. There is no paralysis, but considerable general debility. His bowels are regular, and the tongue clean, but the appetite is impaired; the urine is clear, moderate in quantity, and not albuminous. Heart-sounds normal, but the impulse feeble.

Pulse 80, small and weak.

Nov. 10th. ℞ Quinæ Disulph. gr. iss.
Aquæ distill. ʒj.
Syr. Rhœodos, ʒss.
Acid. Sulph. dil. ɱv.
Ft. Haustus ter die s., Vin. Alb. ʒiv.

With these medicines and middle diet he continued with no appreciable change until the 24th, when he had a kind of fainting fit upon rising to have his bed made, contrary to an order that he should keep in the recumbent posture. This day his diet was changed to milk at his own request. He has been once or twice sick after taking his food.

28th. The sickness has continued, and he to-day has a troublesome hiccough, for which he was ordered

Jul. Ammon., p. r. n.

29th. He has had little sleep, the hiccough, unrelieved by the Julep. Ammon., annoying him much. Dr. Barlow, who now took the ward, ordered him

Æther. Chlor. ɱv.
Vini Opii, ɱv.
ex Mist. Camph. t. d. s.

30th. He is to-day about the same. Has been sick this morning, the vomited matter consisting of food and drink, the hiccough occasionally ceasing.

Dec. 1st. Hiccough still very harassing.

℞ Vini Opii ɱx.
Tinct. Castorei ɱx.
ex Julep. Pimentæ p. r. n.

This was found to relieve the hiccough somewhat.

2nd. He seems considerably weaker, and upon approaching him, his eyelids half closed, allowed the lower sclerotic of the raised eyeballs to be seen. The tongue was moist and clean, and pulse 80, very weak. On speaking to him he roused up and appeared quite as usual, but soon relapsed into the torpid state again. His blood under the quarter inch object-glass presented from forty to sixty white corpuscles in each field, mostly scattered about, but some in patches of two or three and six or eight together.

3rd. Slept better, although the hiccough did not cease. He complains of a constricting pain about the waist; he is tender on pressure over the spleen, where no tumour is to be felt. The tongue to-day is dry, and beginning to be sordid; teeth dirty; pulse week. He presents the same typhoid appearances.

4th. Pulse weaker, dicrotic 96; roused from the torpid state with more difficulty than yesterday. He talks very sensibly, but his wife, who watches by his bedside, states that he wanders in the night.

Jul. Ammon. c. Tinct. Castorei ♏v. p. r. n.

The blood presented the same appearances under the microscope as before.

5th. Hiccough continues, is more feeble, pulse scarcely perceptible, lies in a torpid and typhoid state. When roused, said he was sore all over the body. Tongue and teeth sordid.

6th. Died quietly at 5 a.m.

Sectio-cadaveris.—Nine and a half hours after death in cold wet weather. Rigor mortis, but no decomposition. There was not much emaciation, and the axillæ were slightly discoloured. The countenance was paler than in life, but presented the same olive hue, with the dark patches on the face, forehead, and lips. There was a psoas abscess on the right side, extending from Poupart's ligament to the diseased vertebræ, and holding about a pint of flaky pus. The disease was between the first and the second vertebræ, commencing in the cartilage, and nearly destroying the neighbouring vertebræ at their centres.

The bone surrounding the cavity was red, soft, and infiltrated with strumous matter.

Pleura and bronchi healthy.

Both lungs contained hard masses of grey strumous pneumonic deposit, mostly in the apices, but also in the lower lobes; these masses presented the appearance of a conglomeration of tubercles, held together by inflammatory matter. Heart and pericardium healthy. Heart's weight seven ounces and a half.

The blood on microscopic examination contained the same excess of white corpuscles observed in life. Stomach healthy, slightly adherent to the left supra-renal capsule; its structure was not affected. Spleen large, firm, seven ounces and a half in weight.

Corpuscles visible. The pancreas and all other abdominal organs were healthy. The head was not examined.

Each supra-renal capsule was completely destroyed and converted into a mass of strumous disease, the latter of all degrees of consistency. The left supra-renal capsule had formed at the upper part a close connection with the outer coat of the stomach. The upper part of this capsule seemed fluid, and of the colour of pus; the lower firmer, and of the consistency of putty. The right capsule had all degrees of consistency from the bottom to the top; the lower part almost fluid and resembling pus, the centre putty-like, and above this the matter could be detached in flakes; and at the top it was quite earthy, separate angular pieces being easily detached. Although this patient was known to be suffering under a serious affection of the spine, the ordinary indications of the disease of the supra-renal capsules were sufficiently prominent to justify the prediction, which was so satisfactorily confirmed by the post-mortem examination. It is also worthy of remark that, although the patient, as usual, suffered considerably from irritability of stomach, there was but little change observable in that organ after death.

CASE 4.—Reported by the Ward Clerk.

John Iveson, æt. 22, admitted into Guy's Hospital, March 20, 1854, and died the following day. A stonemason, residing at Lambeth. Last winter he had pain in the stomach and vomiting. He slightly improved, but the day after Christmas was confined to his bed with great pain and vomiting; he vomited matter consisting of a watery fluid. At that time he had "tic douloureux." On admission, his extremities were cold, he was almost pulseless; his hands were blue: he had not had any diarrhœa; he had slight pain, or rather soreness in the hypogastric region; he was quite sensible; the pupils were much dilated. He rallied a little after his admission; had no purging, but vomited bilious matter; had no diabetes or albuminuri. He appeared to die from syncope.

Sectio-cadaveris.—Seventeen hours after death; weather cold; limbs rigid, body tolerably nourished, face of a dingy colour, also the axillæ and hands. Abdomen not distended.

Head.—The dura mater and sinuses were found to be healthy, the membranes injected, and the veins full. There was slight subarachnoid effusion. The grey matter of the cerebrum was rather deep in colour. The brain was in other respects normal.

Chest.—Trachea granular and congested. The right pleura adherent at the posterior and lower parts; on the left side there were firm adhesions at the apex; the bronchi granular; the left apex was a little puckered, and presented several lobules, with iron-grey consolidation and calcareous deposit. The right lung was healthy, with the exception of a single iron-grey consolidation at the apex. The bronchial and mediastinal glands were healthy.

Heart.—Pericardium healthy. There was a white patch on the right ventricle. The right side of the heart was moderately distended with clot, the left entirely and firmly contracted. The valves were healthy, and the muscular fibre, though flaccid, appeared healthy. No fat was found about the heart. Weight seven ounces.

Abdomen.—Peritoneum healthy—viscera moderately contracted. Stomach not distended; at the cardiac extremity there was postmortem solution of the mucous membrane; towards the lesser curvature it was granular, in some parts destroyed, ulcerated; quite superficially there was arborescent injection. On microscopical examination, mucus- and granule-cells were observed. Brunner's glands were very prominent. Ileum with much mucous congestion. Peyer's and solitary glands very distinct, but only hypertrophied. The mesenteric glands were enlarged, firm, and white, full of nuclei, hypertrophied.

Large intestines were healthy.

Liver was of normal form and condition; there was a small amount of fat in the cells; weight two pounds fourteen ounces, containing no arsenic. Gall-bladder healthy; ducts free, but not enlarged. Spleen enlarged—weight six ounces. Pancreas was healthy.

The two supra-renal capsules together weighed forty-nine grains; they appeared exceedingly small and atrophied; the right one was natural, firm; the left deformed by contraction; each adherent to surrounding parts by dense areolar tissue; the section gave a pale and homogeneous aspect; it presented a fibrous tissue, fat, and cells about the size of white blood-corpuscles. The lumbar glands were enlarged. The kidneys coarse, weighing ten ounces. The bladder and prostate were healthy.

The history of this man's case renders it probable that his disease commenced several months prior to his admission into the hospital; and it is not a little remarkable that his earliest complaint was of sickness, vomiting, and pain in the region of the stomach—symptoms which have constituted a more or less prominent feature in every case that has fallen under my notice, and which, in the present instance, were so urgent as to suggest a suspicion of some acrid poison having been received into the stomach.

How far these gastric symptoms, when present, are referable to sympathy existing between diseased capsules and the stomach—how far they depend upon disturbed circulation within the head—how far they are attributable to accidental or essential gastric inflammation—and how far the inflammatory aspect of the gastric mucous membrane is the mere result of severe and repeated vomiting—a more extended observation will probaby determine hereafter.

It was from the presence of these gastric symptoms, the extreme and peculiar prostration of the patient's strength, the great feebleness and smallness of the pulse, the anæmiated eye, the absence of any discoverable lesion to account for the patient's condition, and more especially the dingy discoloration of the face, that led before death to a belief that we should on post-mortem examination find disease of the supra-renal capsules.

It is, moreover, of some significance and importance to observe that, in the present instance, the diseased condition of the supra-renal capsules did not result, as usual, from a deposit either of a strumous or malignant character, but appears rather to have been occasioned by an actual inflammation, that inflammation having destroyed the integrity of the organs, and finally led to their contraction and atrophy.

Case 5.—The following, taken from Dr. Bright's reports of medical cases, presents, according to my belief, a very good illustration of the disease under consideration, and is headed: "Serous Effusion under the Arachnoid and into the Ventricles in a Case of Emaciation, with Bilious Vomiting and diseased Renal Capsules."—Ann Roots was admitted in July, 1829, under one of the surgeons, into Guy's Hospital, on account of a tumour in the left breast and a swelling in the right parotid; but as it was perceived that she was greatly emaciated and apparently sinking, and therefore quite unfit to undergo any operation, she was transferred to the care of the physician. *Her complexion was very dark*—her whole person emaciated; she had no cough, and neither tension nor tenderness of abdomen; she had great difficulty in opening her jaw, owing to the glandular swelling, and could not protrude her tongue. There was no indication but to support the strength. Her stomach soon became irritable; she had bilious vomiting, which reduced her strength, and for a day or two before her death, which took place on the 18th of August, she became drowsy, yet capable of being roused, complaining of some pain over the forehead, and occasionally wandering a little in her intellects.

In the absence of all positive symptoms, I concluded that it was possible some glandular disease, similar to that which had shown itself below the mammæ and under the jaw, might exist internally, giving rise to emaciation and vomiting; and it appeared probable that serous effusion had been going on in the head for the last few days.

Sectio-cadaveris.—Considerable emaciation; and on removing the integuments the scalpel opened into an abscess, containing an ounce or two of pus, situated beneath the mamma of the left side. The dura mater was firmly attached to the skull at the vertex, where the bone was remarkably thin and indented by the glandulæ Pacchioni, and the ordinary opaque deposit which surrounds them. On raising the dura mater several small opacities were observable on the arachnoid; and a very considerable quantity of serous fluid was effused under the arachnoid, raising it into bladders, as well as filling up the hollows between the convolutions. The whole brain was soft and watery, and many vessels showed themselves where horizontal sections were made. In the ventricles about half an ounce of fluid was collected. The choroid plexus was quite exsanguine. Slight adhesions of the pleura pulmonalis and pleura costalis were found, but not sufficient to prevent the lungs from collapsing pretty completely when the air was admitted into the chest. The upper lobe of each lung was in an unhealthy state, looking puckered, and containing one or two masses of earthy matter, besides several small incipient tubercles; the greater part of the lungs, however, was in a very healthy condition. Heart small, but healthy. In the abdomen slight old adhesions had taken place in various parts, but they were composed of the finest transparent cellular tissue; even the omentum, which was glued by them to various parts, both of the intestines and the parietes, had lost none of its delicacy and transparency.

The intestines were healthy, but stained with bile; the mucous membrane healthy; the liver healthy and the gall-bladder full of bile; the pancreas healthy and the spleen also, but just between the pancreas and the spleen a few absorbent glands were enlarged. The glands of the mesentery were also slightly enlarged. The only marked disease was in the renal capsules, both of which were enlarged, lobulated, and the seat of morbid deposits, apparently of a scrofulous character; they were at least four times their natural thickness, feeling solid and hard; on the left side one part had gone into suppuration, containing two drachms of yellow pus. The kidneys themselves healthy. The uterus held down by adhesions in the pelvis.

It does not appear that Dr. Bright either entertained a suspicion of the disease of the capsules before death, or was led at any period to associate the colour of the skin with the diseased condition of these organs, although his well-known sagacity induced him to suggest the probable existence of some internal malignant disease. In this, as in most other cases, we have the same remarkable prostration, the usual gastric symptoms, the same absence of any very obvious and adequate cause of the patient's actual condition, together with a discoloration of the skin, sufficiently striking to have arrested Dr. Bright's attention, even during the life of the patient.

CASE 6.—R. H—, Esq., was a member of the bar, somewhere about middle age. I had the satisfaction of attending him in consultation with Dr. Watson and Mr. Barker, when I was informed that he had been getting thin and emaciated during a period of about twelve months. His appearance and symptoms were very remarkable. He was certainly thin, but not strikingly emaciated, and the surface was soft, loose, and supple. He was greatly anæmiated; his eyes were pearly; he complained of extreme languor and faintness; his pulse, contrary to what is usual in capsular disease, was of good size, but exquisitely soft and compressible; the impulse of the heart was feeble, and palpitation or throbbing with scrobicular pulsation was immediately produced by the slightest exertion; without pain, the stomach was extremely irritable, and vomiting was both urgent and distressing. With these symptoms the surface generally presented a dark, dingy aspect, and there were observed, chiefly on the face, neck, and arms, patches of a rather deep chesnut-brown colour; these chesnut-brown patches were of various sizes and shapes, and were associated here and there with others presenting a singularly white or blanched appearance, arising either in consequence of the latter portions of the integument having remained unaffected, and so contrasting with the surrounding discoloration, or, what is more probable, from their having received a less supply of pigment than natural. A patient inquiry and most careful examination failed to elicit any information, or to detect any lesion, sufficient to afford even a plausible explanation of the patient's singular condition. The violent vomiting pointed to organic, perhaps carcinomatous disease of the stomach; nevertheless the general condition and symptoms did not in other respects seem to warrant such a conclusion; and coupling the existing condition and symptoms with the irregular deposition of dark pigment in the skin, a suspicion was entertained that the whole might arise from disease of the supra-renal capsules. To the last, however, considerable doubt prevailed amongst us as to the true nature of the case—chiefly in consequence of the severity and persistence of the vomiting and from the vomited mucous matters having been occasionally tinged with blood.

The patient speedily sank, and the following report of the morbid appearances discovered after death was furnished, I believe, by my distinguished friend, Dr. Hodgkin.

"The morbid specimens consisted of part of the stomach and duodenum—the termination of the small, and the commencement of the large intestines, with the appendix vermiformis, and the renal capsules, with a small portion of the kidney. They were taken from a man rather beyond middle life, who, for a considerable time, had suffered from obstinate derangement of the stomach. "The coats of the stomach taken unitedly did not produce any preternatural thickness, but rather the reverse; yet there might be a little thickening or increased development of the mucous membrane. The peculiarity of its appearance consisted in a spotted character, not very easily described. Near the pylorus it seemed to consist of a very slight degree of that irregularity which Louis has described as the *état mamelonné*, and which appears to be nothing more than the increased development of a natural structure; but in this instance the elevations were smaller in size, and consequently more numerous, though less prominent than those generally seen towards the middle of the stomach, where this appearance is most frequently noticed. Further from the pylorus, in the direction of the smaller curvature, smaller spots were seen more scattered and distant from each other, and apparently consisting of opaque light-coloured matter, within the semi-transparent substance of the mucous membrane itself, which was generally of a faint dusky-reddish colour. It could not be decided whether these spots depended on any glandular apparatus, yet the idea suggested itself that they might be connected with the follicles of Lieberkühn. Immersion under water, with the intention of facilitating the examination with the microscope, rendered these spots less conspicuous. The largest might equal a small pin's head; the smaller ones scarcely a quarter so large. The duodenum appeared healthy. The portion of small and large intestine, of which the next specimen consisted, offered nothing remarkable in texture. The mucous membrane was tinged with the dingy olive green of the fæcal contents, and ileo-colic valve was rather more prominent than usual in the cæcum. The appendix vermiformis was about three inches in length, but much distended, being about an inch in diameter at its commencement, and becoming gradually less towards the free extremity, where it but little exceeded the normal size. Its peritoneal coat was quite healthy; its general thickness was very little increased; its mucous membrane apparently healthy, of greyish colour, from a little black pigment towards the upper part. Its follicular apparatus was nearly or quite imperceptible. It was completely cut off from the interior of the intestine, the mucous membrane forming a cul-de-sac at both extremities, although there was no apparent want of continuity on the exterior, the septum between the two cavities being merely composed of the two mucous membranes united by cellular tissue. No appearance of cicatrix was discovered, indicating that the separation was of long standing, if not congenital. The contents of the appendix consisted principally of a transparent colloid or thick mucoid secretion, partly of a light straw colour, partly tinged with blood. Interspersed through it, but especially towards the upper part, was an opaque white substance, of the same consistence, resembling coagulated milk or ground white lead. A few points were blackened by pigments.

Examined with the microscope, the transparent portion exhibited no determinate structure, but a slight tendency to filamentous arrangement. The whole portion was made up of a congeries of oil-globules, varying in size, but all very minute. The black pigment appeared to pervade some of the oil-globules, rather than itself to compose distinct corpuscles. The basis of this collection was undoubtedly the mucus of the appendix itself, retained by the want of any excretory passage. "The small fragment of kidney appeared to be of healthy structure, but both the renal capsules were enlarged (the united weight of the two being one and a half ounce), of rather irregular surface, and considerably indurated. When cut into, instead of exhibiting the ordinary appearance of combination of dark and yellow substances, they seemed to consist of a firm, slightly transparent reddish basis, interspersed with irregular spots of opaque yellow matter, the whole bearing a strong resemblance to an enlarged mesenteric gland, mottled with tubercular deposit. Such was probably the nature of the change which the organ had undergone. The naked eye could discover no trace of cystiform arrangement, and the opaque matter, when examined with the microscope, exhibited a copious amount of fatty matter, but no nucleated cells."

It was to me a matter of much regret that I had not an opportunity of employing an artist to make an exact representation of the singular discoloration observed upon the skin, and the more so because, although agreeing in general character with those observed in other cases, there was a manifest peculiarity, as well in the intensity as in the mode of distribution of these discolorations. With universal dinginess of the surface there were, especially about the neck, hands, and arms, several well-defined patches of a deeper or somewhat chesnut-brown hue, interspersed here and there with blanched or almost dead-white portions of integument, contrasting in a very remarkable manner with both the general dinginess and deep brown patches; and what is very remarkable, wherever the integument presented the blanched or dead-white appearance, the hairs upon its surface were observed to have turned completely white. The superiority of a coloured drawing over the most elaborate verbal description, in conveying a correct idea of any morbid appearance, is so universally felt and acknowledged, that I have great satisfaction in being now able to furnish one, which may most fairly and faithfully be applied to the above case. Very recently (March, 1855) I was requested to visit a patient (Mr. S—), about sixty years of age, who presented, in a strongly marked degree, the indications of diseased renal capsules. The history, mode of attack, the progress, the anæmia, the extreme feebleness of the heart's action, the uneasiness and irritability of the stomach, and the discoloration of the skin, were all such as characterise the disease generally, and bore the closest resemblance to the above case in particular. My belief was that the capsules were affected with malignant disease, and that probably some other structures about the posterior mediastinum might have been in a similar condition, as the patient had slight œdema of both the upper extremities, whilst the lower limbs remained free. Anxious as I was to procure a post-mortem examination, it was most firmly and peremptorily refused; and it was only through the kind and persevering efforts of my friend, Mr. Parrott, of Clapham, that I succeeded in gaining permission to have a sketch taken of the discoloured integument. Of course, this representation does not carry along with it such authority and conviction as one taken from a subject actually proved to have had diseased capsules. Nevertheless, I entertain no doubt whatever that the capsules were diseased; and even if they were not, I hold myself answerable for the most perfect resemblance between the two cases, so far as the infection of the integument was concerned.

CASE 7.—The following case having been under the care of one of the surgeons for "carcinoma" of the mamma, I have not been able to furnish any record of the symptoms during life. The corpse, however, presented appearances sufficiently striking to arrest the attention, and call forth the correct prediction of Dr. Lloyd, the inspector, who kindly furnished me with the following report.

"*Sectio-cadaveris.*—M. T—, æt. 60, cancerous disease of the mamma, with cancerous disease of the supra-renal capsules. Sixteen hours after death. Body extremely emaciated; the left mamma presented a very extensive ulcerated phagedænic, malignant tumour, occupying the whole of the upper part of the left side of the chest, infiltrating the cellular tissue, the skin and intercostal muscles with carcinomatous material. *The colour of the skin covering the face, arms, and chest was of a peculiar light brown swarthy hue. Chest.*—On raising the sternum and cartilages it was found that the malignant growth had passed through the pleura and invaded the lung on the left side, for a space of the size of the palm of the hand, by direct continuity of structure. The pleural cavity of the side contained about sixteen ounces of dark-coloured fluid. The lower lobe of the left lung was compressed and sank in water. The upper lobe was healthy. The right lung was healthy. *Heart* was small and flabby. *Abdomen.*—The liver was contracted, irregular on its surface, of yellow colour, containing abundance of fat, burning brilliantly in the spirit lamp; upon its surface were several nodules of cancerous development. The gall-bladder was occupied in its entire extent by a calculus, and did not contain any bile. Both supra-renal capsules contained a considerable amount of cancerous deposit, invading their entire structure, and almost obliterating their cavities. The kidneys were contracted and granular. The uterus healthy, but atrophied." I have already expressed my belief that the urgency of the symptoms, and the quick or slow progress of the disease, are determined by the activity or rapidity of the morbid change going on in the capsules, and by the actual amount or degree of that change; and that universal disease of both capsules will, in all probability, be found to prove uniformly fatal. These views appear to be countenanced by the character, progress, and termination of the cases already given, and receive additional confirmation from the history of the following, in which the morbid change was limited to a single capsule, and in which the constitutional and local consequences indicate a corresponding result.

CASE 8.—Reported by the Ward Clerk.

Elizabeth Hannah Lawrence, æt. 53, was admitted into Guy's Hospital under Dr. Babington, March 30, 1853.

Appearance.—A short woman, emaciated and feeble, skin harsh and dry, and of a darkish hue. The folds of the axillæ were remarkably dark; coloured patches, about the size of the palm of the hand, were observed raised in wrinkles, and resembling a slight ichthyosis; also a very dark-brown areola around the umbilicus. Hair grey; much long hair on lips and chin.

Previous history.—Is a single woman, and has always been a servant; has been living of late in Trinity Street, Borough. Was always thin, but yet always enjoyed good health.

Present history.—Four months ago an eruption appeared on her body, for the cure of which she went to the Cutaneous Infirmary at Blackfriars. In a short time she was cured, and just as the eruption disappeared the present stomach symptoms began. For three months she has had vomiting, with pain in the abdomen and back, particularly in the latter. She has thrown up no blood. She was sent to the hospital as a case of malignant disease of the stomach. The stomach can be felt as a hard tumour in the abdomen; no remains of eruption on the skin. The vomiting continued after admission, and in three days she died from exhaustion.

Sectio-cadaveris.—External appearance.—The body that of a small emaciated woman, with a fair skin and dark hair, presenting certain peculiar discolorations. On either side of the neck there was a tawny appearance, which would not have been remarked had it not been for three still more marked tawny patches, one on the centre of the sternum, the other two under either axilla. The skin also, besides presenting this yellowish-brown appearance, was somewhat raised and wrinkled or corrugated. These marks led me to prognosticate disease of the supra-renal capsules before opening the body, believing them to be the marks pointed out by Dr. Addison. *Thorax.*—The lungs were congested, exuding a frothy serum, and easily lacerable. *Heart.*—Small and lacerable. The mediastinal glands in one or two instances carcinomatous. *Abdomen.*—Was shrunk and contracted. *Stomach.*—The walls of the stomach from the pylorus through the lesser curvature were thickened, presenting on the stomach externally a peculiar network appearance, containing a transparent stroma; beneath this another layer with its fibres longitudinally arranged, of strong cellular material; within this, the mucous membrane whole and intact; the entire thickness being about three quarters of an inch at the pylorus, gradually decreasing to a quarter at the commencement of the cardia. The mucous membrane lower down was here and there destroyed by ulceration, and this ulceration in one instance of an eighth of an inch in size. The stomach was contracted and empty; externally to the stomach several of the glands were affected, even to the head of the pancreas, but the pancreas itself was not affected. Several of the lumbar glands were enlarged. The left supra-renal capsule was infiltrated with malignant material, and closely adherent to the vessels of the kidney. The kidney itself was healthy. The uterus contained three fibrous tumours the size of walnuts.

Although this woman only survived four days after her admission into the hospital, we were led by the partial discoloration of the skin to anticipate disease of the capsules, one only of which, however, was found to be implicated. It will have been perceived that in a certain number of the cases already given, either strumous or malignant disease existed in other parts or organs, as well as in the capsules, and of course, in the midst of such complications, there is often more or less difficulty in satisfactorily unravelling the case in all its details during life; nevertheless, as we know that without any such complication whatever, mere disease of the capsules themselves has proved sufficient to produce such alarming symptoms and such serious consequences, it cannot with any show of reason be alleged that these peculiar symptoms, when present, arise exclusively from the accidental complication of other organs.

In the present instance, as in some others, the immediate cause of death, as well as of many of the most distressing symptoms during life, was unquestionably carcinomatous disease of the stomach.

CASE 9.—Thomas Clouston, æt. 58, admitted into Guy's Hospital February 11th, 1852, under Dr. Barlow. A muscular and strong-built man, of a sanguine temperament, and dark complexion. He has been a married man, but his wife died about twenty years ago. His occupation has been that of a sailor, and according to his own statement, he has led a very sober life. His general health has been very good. About five years since he had a hernia in the left inguinal region, for which he has since worn a truss. This has never given him any difficulty to return. About two months ago he came from Liverpool, in which place he had settled, not intending to go to sea again; and was taken on board the "Dreadnought" for stricture. His general health was quite good at this time, but while in the "Dreadnought" he began to lose his appetite and to feel generally unwell; he had likewise some affection of the left eye, in which he is now nearly blind.

On Saturday the 8th, he left the ship at his own request, thinking that he might be better on land; after waiting two or three days he found that he got no better, and his friends advised him to come to the hospital.

Present symptoms.—He complains of a sensation of sickness without actual vomiting; and tightness over the epigastrium. His countenance is anxious. He has no pain in any part. He has rigors followed by mild sweats, every five or six hours, the rigors usually lasting about an hour. The abdomen is tense and tympanitic; not tender to the touch, excepting over the upper part. The liver does not appear enlarged. His chest is broad and well formed; the motion of the ribs moderate, resonant on the percussion; and the lungs are apparently sound. The heart's sounds are normal. Pulse rather feeble, 80. Tongue injected at the tip and edges, coated with a light-brown fur, very dry. Urine of about average quantity, rather large than otherwise; of a high colour, acid, and does not coagulate by heat. The bowels have been regular. After he had been in a few hours, he brought up a lage quantity of beer. Ordered

Mixt. Efferves. 4tis horis.

Feb. 12th.—The sickness has not returned, but he is without any appetite. He slept but little.

13th.—He is much the same; but has a more sallow and sunken expression of countenance. He complains of nothing but loss of appetite and general debility. His tongue continues dry and coated with a brownish fur. His bowels have been relaxed, and passed his motions partly involuntarily.

14th.—No special change.

17th.—He seems rather better; he had a little breakfast, and enjoyed it.

18th.—He has relapsed into his former state, having no appetite, and complaining of great debility and thirst. He has ʒiv of sherry daily.

20th.—There is but little change in him, his *countenance appears to grow darker*, and his strength seems gradually failing. His bowels are rather irritable. Ordered

Enema Amyli c. Syr. ʒss;
Inf. Cuspariæ, iss, t. d.

25th.—He has been getting gradually weaker, without showing any special symptoms in addition to those mentioned. He died this morning.

Sectio-cadaveris.—None was allowed beyond the brain and abdomen. Of the former there was considerable softening, and a large amount of subarachnoid fluid. The kidneys were slightly enlarged, mottled, and in some parts the cortical substance was entirely degenerated into fat. A few tubercles were observed on the surface. The tunic was very easily taken from the surface; tubercles were also observed in the spleen, and on the peritoneum covering the termination of the ileum. Tubercular deposit was likewise found in one of the supra-renal capsules.

The development of tubercles on various parts, as well as in one of the supra-renal capsules, sufficiently attests the strumous character of the patient's disease; and it is difficult to divest oneself of the notion that the disease in the supra-renal capsule had some share in producing the peculiar symptoms which immediately preceded the fatal result, whatever importance may be attached to the state of the kidneys and cerebral complications. At all events, the discoloration of the skin indicated before death the existence of capsular disease; and it is worthy of remark that in this instance the deposition of pigment-cells was not limited to the integument, but was found scattered in small masses over the omentum, the mesentery, and the cellular tissue on the interior of the abdominal parietes.

CASE 10.—Jane Roff, æt. 28. This person was admitted into the Obstetric Ward, labouring under cancer of the uterus, February 4th, 1852. She died February 8th, and on the 9th the body was placed on the table for inspection. When proceeding to perform this duty, Dr. Lloyd was struck with the peculiar dingy appearance of the skin, and in consequence, prior to commencing, sought me to look at it. The appearance, though not very strongly marked, was certainly such as to create a strong suspicion that something was wrong with the capsules. On exposing the organ on the right side it presented a perfectly healthy appearance, and we felt disposed to conclude that our anticipation would turn out to be erroneous. On proceeding to examine the left capsule, however, we were much surprised to find a very extraordinary, and, I suspect, an extremely rare condition of parts. A malignant tubercle had been developed at that precise point where the large vein escapes from the organ; this tubercle projected into the interior of the vein, so as almost or entirely to obstruct it, and had, moreover, led to rupture and effusion into or a sort of apoplexy of the capsule itself. This case would render it probable that the excess of dark pigment, so characteristic of renal capsular disease, depended rather upon an interruption of some special function than upon the nature of the organic change; for, with the exception of the manifestly recent sanguineous effusion into its tissue, the capsule itself did not appear to have undergone any considerable deterioration.

CASE 11.—I may observe, in conclusion, that very recently there was examined, at Guy's Hospital, the body of a person—William

Godfrey—who had died of cancer, affecting the thoracic parietes, and extending through the lungs. Quite unexpectedly there was found extensive disease of one of the supra-renal capsules, the organ being very much enlarged, and converted into a hard mass of apparently carcinomatous disease. On referring to the notes of the case, as taken by the clinical clerk, I found it stated that *the patient's face presented a dingy hue,* although he was naturally of a fair complexion, with reddish or sandy hair in the pubes; and, moreover, the face of the corpse was ascertained to present a freckled and dingy appearance, with a slight brown discoloration at the root of the nose and at each angle of the lips.

THE END.

PRINTED BY J. E. ADLARD, BARTHOLOMEW CLOSE.

INDEX.

Abdominal friction sound, 89
Abscess of lung following pneumonia, 2, 23
Addison's disease of capsules, 211
— keloid, 177
Adelon, on pulmonary vein, 3
Air-cells, serous character of, 20
Albuminization in inflammation, 22
Albuminous induration of lung, 27
Alibert's keloid, 158, 170
Amaurosis, with hysterical paralysis, 204
Anasarca, with fatty liver, 105
Anatomy of air-cells, 19
— of the lungs, 1
Apoplexy of lung leading to gangrene, 3

Babington, Dr., case of Addison's disease, 234
Barlow, Dr., case of Addison's disease, 236
Bateman, on vitiligo, 157
Bichat, on anatomy of pulmonary vein, 3
Bile, analysis of, in fatty degeneration of liver, 100
Bird, Dr. Golding, analysis of bile in fatty liver, 100
Brain disease simulated by pneumonia, 14
— disorders of, with kidney disease, 187
Bright, Dr., case of Addison's disease, 228
Bronchial tubes, dilatation of, 44
— mistaken for tubercle, 61

Capsules, supra-renal, disease of, 211
Carnification of lung (Laennec), 25
Chest, physical examination in diseases of, 65
Chorea, use of electricity in, 202, 206, 207
Cloquet, on anatomy of pulmonary vein, 3

Convulsive diseases, use of electricity in, 195
Cough in pneumonia dependent on bronchitis, 15
Cyst, ovarian, accidental rupture of, 15

Dieburg, Dr., on keloid, 175
Dropsy, ovarian, with rupture of cyst, 155
— with fatty liver, 105

Electricity, use of, in convulsive diseases, 195

Fatty degeneration of the liver, 99
Females, disorders of, connected with uterine irritation, 109
Fever, continued, simulated by pneumonia, 13

Gull, Dr., case of Addison's disease, 220

Hæmoptysis, slight, importance of, 72
Hodgkin, Dr., case of Addison's disease, 230
Hysterical paralysis, use of electricity in, 203

Inflammation, albuminizing tendency of, 22
Iodine, use of, in true keloid, 179

Jaundice, with vitiligoidea, 163

Keloid, of Alibert, 158, 170
— spurious, 173
— true, use of iodine in, 177, 179
Kidney disease, brain disorders with, 187

Lebert, on keloid, 176
Liver, derangement of, with vitiligoidea, 163
— fatty degeneration of, 99
— anasarca with, 105
— condition of skin with disease of, 101

Lloyd, Dr., case of Addison's disease, 238
Lung, abscess of, as result of pneumonia, 2
— air-cells of, anatomy of, 19
— albuminous induration of, 27
— anatomy of the, 1
— apoplexy of, a cause of gangrene, 3
— changes in apex of, 37
— carnification of, 25
— circumscribed gangrene of, 23
— diffuse gangrene of, 23
— granular induration of, 27
— grey hepatization of, 22
— grey induration of, 28
— œdema of, seat of, 21
— red hepatization of, 21
— tubercle of, analogous to that of serous membranes, 30
— uniform albuminous induration of, 35

Meckel, on anatomy of pulmonary vein, 3

Œdema of lungs, seat of, 21
Ollivier, on anatomy of pulmonary vein, 3
Ovarian dropsy, case of, with rupture of cyst, 155

Paralysis, hysterical, use of electricity in, 203
Phthisis, pathology of, 39
— pneumonic, 48
— tubercular, 56
— tuberculo-pneumonic, 49
Pneumonia and its consequences, 17
— absence of pain in, 16
— cough in, dependent on bronchitis, 15
— diagnosis of, 7
— effects of, 19
— giving rise to abscess, 2
— identical with inflammatory tubercle, 2
— lobular, 24
— ordinary changes in, 33

Pneumonia, permanent effects of, 27
— rarely terminates in abscess, 23
— seat of, 8, 18
— simulating brain disease, 14
— simulating fever, 13
— simulating hydrocephalus, 14
— suffocative, 54
Pneumonic phthisis, 48
Preface to papers on diseases of lung, xix
Pulmonary tubercle analogous to that of serous membranes, 30
— artery, anatomy of, 2
— vein, anatomy of, 3

Rayer, on keloid, 158
Reissessen, on anatomy of air-cells, 19
Renal capsules, supra-, diseases of, 211

Serous character of air-cells, 20
Skin, bronzing of, in Addison's disease, 215
— condition of, with fatty liver, 101
— keloid of, 169
— vitiligoidea of, 157
Supra-renal capsules, disease of, 211

Tubercle, asthenic, 50
— compound, 50
— inflammatory, identical with pneumonia, 2, 28
— of serous membranes analogous to pulmonary, 30
— sthenic, 49
— with fatty liver, 102
Tubercular phthisis, 56
Tuberculo-pneumonic phthisis, 49

Uterine irritation, disorders connected with, 109

Vein pulmonary, anatomy of, 3
Vitiligo (Willan), 157
Vitiligoidea plana, 157
— tuberosa, 157

Willan, description of vitiligo, 157

ERRATA.

Page 24, 12th line from top, *for* "these" *read* "those."
„ 99, 5th line from top, *for* "accept" *read* "except."
„ 159, 17th line from top, *for* "hypochondria" *read* "hypochondrium."
„ 170, 2nd line from bottom, *for* "from far as" *read* "far from."